THE ANALYSIS OF MATTER

BERTRAND RUSSELL

Available from Spokesman

Philosophical Writings

The Analysis of Matter
The Analysis of Mind
The Foundations of Geometry
An Inquiry into Meaning and Truth
Introduction to Mathematical Philosophy
Logic and Knowledge
My Philosophical Development
Mysticism and Logic
The Philosophy of Leibniz
Political Ideals

* * *

Other Titles

German Social Democracy
Has Man a Future
Icarus or the Future of Science
Justice in War-Time
Let the People Think
The Life of Bertrand Russell in Pictures and his own Words
Nightmares of Eminent Persons
Portraits from Memory
The Practice and Theory of Bolshevism
The Problem of China
Roads to Freedom
Satan in the Suburbs

BERTRAND RUSSELL

The Analysis of Matter

SPOKESMAN

First published in 1927
This edition published in 2007 by
Spokesman
Russell House
Bulwell Lane, Nottingham NG6 0BT, England
Phone 0115 9708318
Fax 0115 9420433
e-mail elfeuro@compuserve.com
www.spokesmanbooks.com

ISBN10: 0 85124 740 7
ISBN13: 978-0-85124-740-3

A CIP Catalogue record is available from the British Library

Printed by the Russell Press Ltd.
(phone 0115 9784505 www.russellpress.com)

CONTENTS

CONTENTS

PART III

THE STRUCTURE OF THE PHYSICAL WORLD

PREFACE

THE attempt to discover the philosophical outcome of modern physics is one which, at the present moment, is beset with great difficulties. For, while the theory of relativity has achieved, at least temporarily, a stable form, the theory of quanta and of atomic structure is developing with such rapidity that it is impossible to guess what form it will take a few years hence. In these circumstances, it is necessary to exercise judgment as to the parts of the theory which are definitively established and the parts which are likely to be modified in the near future. For one who, like the present author, is not a professional physicist, the exercise of such judgment is difficult, and is likely to be occasionally at fault. The subject of the relation of " matter " to what exists, and generally of the interpretation of physics in terms of what exists, is, however, not one of physics alone. Psychology, physiology, mathematical logic, and philosophy are all required, in addition to physics, for the adequate discussion of the theme with which this volume deals. Consequently certain shortcomings on the part of a single author, however regrettable they may be, are perhaps scarcely avoidable.

I am indebted to Mr R. H. Fowler, F.R.S., Mr M. H. A. Newman of St. John's College, Cambridge, and Mr F. P. Ramsey of King's College, Cambridge, for valuable help in regard to certain portions of the work; also to Dr D. M. Wrinch for kindly reading the whole in typescript and supplying many valuable criticisms and suggestions.

Certain portions of the book were delivered as the Tarner Lectures in Trinity College, Cambridge, during the Michaelmas Term, 1926. The book was, however, in preparation before the invitation to give these lectures was received, and contains a good deal of material for which there seemed no place in the lectures.

Since the purpose of the book is philosophical, it has been my endeavour to avoid physical and mathematical technicalities as far as possible. Some modern doctrines, however, perhaps because they are still recent, I have not succeeded in translating into non-mathematical language. In regard to them, I must beg the indulgence of the non-mathematical reader if he finds too many symbols, and of the mathematical reader if he finds too few.

B. R.

January, 1927.

THE ANALYSIS OF MATTER

CHAPTER I

THE NATURE OF THE PROBLEM

APART from pure mathematics, the most advanced of the sciences is physics. Certain parts of theoretical physics have reached the point which makes it possible to exhibit a logical chain from certain assumed premisses to consequences apparently very remote, by means of purely mathematical deductions. This is true especially of everything that belongs to the general theory of relativity. It cannot be said that physics as a whole has yet reached this stage, since quantum phenomena, and the existence of electrons and protons, remain, for the moment, brute facts. But perhaps this state of affairs will not last long; it is not chimerical to hope that a unified treatment of the whole of physics may be possible before many years have passed.

In spite, however, of the extraordinary successes of physics considered as a science, the philosophical outcome is much less clear than it seemed to be when less was known. The purpose of the present chapter is to discuss what is meant by the " philosophical outcome " of physics, and what methods exist for determining its nature.

There are three kinds of questions which we may ask concerning physics or, indeed, concerning any science. The first is: What is its logical structure, considered as a deductive system? What ways exist of defining the entities of physics and deducing the propositions from an initial apparatus of entities and propositions? This is a problem in pure mathematics, for which, in its fundamental portions, mathematical logic is the proper instrument. It is not quite correct to speak, as we did just now, of "initial entities and propositions." What we really have to begin with, in this treatment, is hypotheses containing variables. In geometry, this procedure has become familiar. Instead of "axioms," supposed to be "true," we have the hypothesis that a set of entities (otherwise undefined) has certain enumerated properties. We proceed to prove that such a set of entities has the properties which constitute the propositions of Euclidean geometry, or of whatever other geometry may be occupying our attention. Generally it will be possible to choose many different sets of initial hypotheses which will all yield the same body of propositions; the choice between these sets is logically irrelevant, and can be guided only by æsthetic considerations. There is, however, considerable utility in the discovery of a few simple hypotheses which will yield the whole of some deductive system, since it enables us to know what tests are necessary and sufficient in deciding whether some given set of entities satisfies the deductive system. Moreover, the word "entities," which we have been using, is too narrow if used with any metaphysical implication. The "entities" concerned may, in a given application of a deductive system, be complicated logical structures. Of this we have examples in pure mathematics in the definitions of cardinal numbers, ratios, real numbers, etc. We must be prepared for the possibility of a similar result in physics, in the definition of a "point" of space-time, and even in the definition of an electron or a proton.

The logical analysis of a deductive system is not such a definite and limited undertaking as it appears at first sight. This is due to the circumstance just mentioned—namely, that what we took at first as primitive entities may be replaced by complicated logical structures. As this circumstance has an important bearing upon the philosophy of physics, it will be worth while to illustrate its effect by examples from other fields.

One of the best examples is the theory of finite integers. Weierstrass and others had shown that the whole of analysis was reducible to propositions about finite integers, when Peano showed that these propositions were all deducible from five initial propositions involving three undefined ideas.* The five initial propositions might be regarded as assigning certain properties to the group of three undefined ideas, the properties in question being of a logical, not specifically arithmetical, character. What was proved by Peano was this: Given any triad having the five properties in question, every proposition of arithmetic and analysis is true of this triad, provided the interpretation appropriate to this triad is adopted. But it appeared further that there is one such triad corresponding to each infinite series $x_1, x_2, x_3, \ldots x_n, \ldots$, in which there is just one term corresponding to each finite integer. Such series can be defined without mentioning integers. Any such series could be taken, instead of the series of finite integers, as the basis of arithmetic and analysis. Every proposition of arithmetic and analysis will remain true for any such series, but for each series it will be a different proposition from what it is for any other series.

Take, in illustration, some simple proposition of arithmetic, say: "The sum of the first n odd numbers is n^2." Suppose we wish to interpret this proposition as applying to the progression $x_0, x_1, x_2, \ldots x_n, \ldots$ In this progression, let R be the relation of each term to its successor. Then "odd numbers" will mean "terms having to x_1 a relation which is a power of R^2," where R^2 is the relation of an x to the next x but one.† We can now define R^{x_n} as meaning that power of R which relates x_0 to x_n, and we can further define $x_m + x_n$ as meaning

* On this subject, cf. *Principles of Mathematics*, chap. xiv.

† The definition of powers of a relation, in a form not involving numbers, is set forth in *Principia Mathematica*, *91.

that x to which x_m has the relation R^{x_n}. This decides the interpretation of "the sum of the first n odd numbers." To define n^2, it will be best to define multiplication. We have defined R^{x_n}; consider the relation formed by the relative product of the converse of R together with R^{x_n}. This relation relates x_1 to x_n; its square relates x_2 to x_{2n}; its cube relates x_3 to x_{3n}, etc. Any power of this relation can be shown to be equivalent to a certain power of the converse of R multiplied relatively by a certain power of R^{x_n}. There is thus one power of this relation which is equivalent to moving backward from x_m to x_0, and then forward; the term to which the forward movement takes us is defined as $x_m \times x_n$. Thus we can now interpret x_n^2. It will be found that the proposition from which we started is true with this interpretation.

It follows from the above that, if we start from Peano's undefined ideas and initial propositions, arithmetic and analysis are not concerned with definite logical objects called numbers, but with the terms of any progression. We may call the terms of *any* progression 0, 1, 2, 3, . . ., in which case, with a suitable interpretation of $+$ and $\times$, all the propositions of arithmetic will be true of these terms. Thus 0, 1, 2, 3, . . ., become "variables." To make them constants, we must choose some one definite progression; the natural one to choose is the progression of finite cardinal numbers as defined by Frege. What were, in Peano's methods, primitive terms are thus replaced by logical structures, concerning which it is necessary to prove that they satisfy Peano's five primitive propositions. This process is essential in connecting arithmetic with pure logic. We shall find that a process similar in some respects, though very different in others, is required for connecting physics with perception.

The general process of which the above is an instance will be called the process of "interpretation." It frequently happens that we have a deductive mathematical system, starting from hypotheses concerning undefined objects, and that we have reason to believe that there are objects fulfilling these hypotheses, although, initially, we are unable to point out any such objects with certainty. Usually, in such cases, although many different sets of objects are abstractly available as fulfilling the hypotheses, there is one such set which is much more important than the others. In the above instance, this set was the cardinal numbers. The substitution of such a set for the undefined objects is "interpretation." This process is essential in discovering the philosophical import of physics.

The difference between an important and an unimportant interpretation may be made clear by the case of geometry. Any geometry, Euclidean or non-Euclidean, in which every point has co-ordinates which are real numbers, can be interpreted as applying to a system of sets of real numbers—*i.e.* a point can be taken to *be* the series of its co-ordinates. This interpretation is legitimate, and is convenient when we are studying geometry as a branch of pure mathematics. But it is not the *important* interpretation. Geometry is important, unlike arithmetic and analysis, because it can be interpreted so as to be part of applied mathematics—in fact, so as to be part of physics. It is this interpretation which is the really interesting one, and we cannot therefore rest content with the interpretation which makes geometry part of the study of real numbers, and so, ultimately, part of the study of finite integers. Geometry, as we shall consider it in the present work, will be always treated as part of physics, and will be regarded as dealing with objects which are not either mere variables or definable in purely logical terms. We shall not regard a geometry as satisfactorily interpreted until its initial objects have been defined in terms of entities forming part of the empirical world, as opposed to the world of logical necessity. It is, of course, possible, and even likely, that various different geometries, which would be incompatible if applied to the same set of objects, may all be applicable to the empirical world by means of different interpretations.

So far, we have been considering the logical analysis of physics, which will form the topic of Part I. But in relation to the interpretation of geometry we have already been brought into contact with a very different problem—namely, that of the application of physics to the empirical world. This is, of course, the vital problem: although physics can be pursued as pure mathematics, it is not as pure mathematics that physics is important. What is to be said about the logical analysis of physics is therefore only a necessary preliminary to our main theme. The laws of physics are believed to be at least approximately true, although they are not logically necessary; the evidence for them is empirical. All empirical evidence consists, in the last analysis, of perceptions; thus the world of physics must be, in some sense, continuous with the world of our perceptions, since it is the latter which supplies the evidence for the laws of physics. In the time of Galileo, this fact did not seem to raise any very difficult problems, since the world of physics had not yet become so abstract and remote as subsequent research has made it. But already in the philosophy of Descartes the modern problem is implicit, and with Berkeley it becomes explicit. The problem arises because the world of physics is, *prima facie*, so different from the world of perception that it is difficult to see how the one can afford evidence for the other; moreover, physics and physiology themselves seem to give grounds for supposing that perception cannot give very accurate information as to the external world, and thus weaken the props upon which they are built.

This difficulty has led, especially in the works of Dr Whitehead, to a new interpretation of physics, which is to make the world of matter less remote from the world of our experience. The principles which inspire Dr Whitehead's work appear to me essential to a right solution of the problem, although in the detail I should sometimes incline to a somewhat more conservative attitude. We may state the problem abstractly as follows:

The evidence for the truth of physics is that perceptions occur as the laws of physics would lead us to expect—*e.g.* we see an eclipse when the astronomers say there will be an eclipse. But physics itself never says anything about perceptions; it does not say that we shall see an eclipse, but says something about the sun and moon. The passage from what physics asserts to the expected perception is left vague and casual; it has none of the mathematical precision belonging to physics itself. We must therefore find an interpretation of physics which gives a due place to perceptions; if not, we have no right to appeal to the empirical evidence.

This problem has two parts: to assimilate the physical world to the world of perceptions, and to assimilate the world of perceptions to the physical world. Physics must be interpreted in a way which tends towards idealism, and perception in a way which tends towards materialism. I believe that matter is less material, and mind less mental, than is commonly supposed, and that, when this is realized, the difficulties raised by Berkeley largely disappear. Some of the difficulties raised by Hume, it is true, have not yet been disposed of; but they concern scientific method in general, more particularly induction. On these matters I do not propose to say anything in the present volume, which will throughout assume the general validity of scientific method properly conducted.

The problems which arise in attempting to bridge the gulf between physics (as commonly interpreted) and perception are of two kinds. There is first the epistemological problem: what facts and entities do we know of that are relevant to physics, and may serve as its empirical foundation? This demands a discussion of what, exactly, is to be learnt from a perception, and also of the generally assumed physical causation of perceptions—*e.g.* by light-waves or sound-waves. In connection with this latter question, it is necessary to consider how far, and in what way, a perception can be supposed to resemble its external cause, or, at least, to allow inferences as to characteristics of that cause. This, in turn, demands a careful consideration of causal laws, which, however, is in any case a necessary part of the philosophical analysis of physics. Throughout this inquiry, we are asking ourselves what grounds exist for supposing that physics is " true." But the meaning of this question requires some elucidation in connection with what has already been said about interpretation.

Apart altogether from the general philosophical problem of the meaning of " truth," there is a certain degree of vagueness about the question whether physics is " true." In the narrowest sense, we may say that physics is " true " if we have the perceptions which it leads us to expect. In this sense, a solipsist might say that physics is true; for, although he would suppose that the sun and moon, for instance, are merely certain series of perceptions of his own, yet these perceptions could be foreseen by assuming the generally received laws of astronomy. So, for example, Leibniz says:

" Although the whole of this life were said to be nothing but a dream, and the visible world nothing but a phantasm, I should call this dream or phantasm real enough, if, using reason well, we were never deceived by it."*

A man who, without being a solipsist, believes that whatever is real is mental, need have no difficulty in declaring that physics is " true " in the above sense, and may even go further, and allow the truth of physics in a much wider sense. This wider sense, which I regard as the more important, is as follows: Given physics as a deductive system, derived from certain hypotheses as to undefined terms, do there exist particulars,

* *Philosophische Werke*, Gerhardt's edition, vol. vii., p. 320.

or logical structures composed of particulars, which satisfy these hypotheses? If the answer is in the affirmative, then physics is completely "true." We shall find, if I am not mistaken, that no conclusive reason can be given for a fully affirmative answer, but that such an answer emerges naturally if we adopt the view that all our perceptions are causally related to antecedents which may not be perceptions. This is the view of common sense, and has always been, at least in practice, the view of physicists. We start, in physics, with a vague mass of common-sense beliefs, which we can subject to progressive refinements without destroying the truth of physics (in our present sense of "truth"); but if we attempt, like Descartes, to doubt all common-sense beliefs, we shall be unable to demonstrate that any absurdity results from the rejection of the above hypothesis as to the causes of perceptions, and we shall therefore be left uncertain as to whether physics is fully "true" or not. In these circumstances, it would seem to be a matter of individual taste whether we adopt or reject what may be called the realist hypothesis.

The epistemological problem, which we have just been stating in outline, will occupy Part II. of the present work. Part III. will be occupied with the outcome for ontology—*i.e.* with the question: What are the ultimate existents in terms of which physics is true (assuming that there are such)? And what is their general structure? And what are the relations of space-time, causality, and qualitative series respectively? (By "qualitative series" I mean such as are formed by the colours of the rainbow, or by notes of various pitches.) We shall find, if I am not mistaken, that the objects which are mathematically primitive in physics, such as electrons, protons, and points in space-time, are all logically complex structures composed of entities which are metaphysically more primitive, which may be conveniently called "events." It is a matter for mathematical logic to show how to construct, out of these, the objects required by the mathematical physicist. It belongs also to this part of our subject to inquire whether there is anything in the known world that is not part of this metaphysically primitive material of physics. Here we derive great assistance from our earlier epistemological inquiries, since these enable us to see how physics and psychology can be included in one science, more concrete than the former and more comprehensive than the latter. Physics, in itself, is exceedingly abstract, and reveals only certain mathematical characteristics of the material with which it deals. It does not tell us anything as to the intrinsic character of this material. Psychology is preferable in this respect, but is not causally autonomous: if we assume that psychical events are subject, completely, to causal laws, we are compelled to postulate apparently extra-psychical causes for some of them. But by bringing physics and perception together, we are able to include psychical events in the material of physics, and to give to physics the greater concreteness which results from our more intimate acquaintance with the subject-matter of our own experience. To show that the traditional separation between physics and psychology, mind and matter, is not metaphysically defensible, will be one of the purposes of this work; but the two will be brought together, not by subordinating either to the other, but by displaying each as a logical structure composed of what, following Dr H. M. Sheffer,* we shall call "neutral stuff." We shall not contend that there are demonstrative grounds in favour of this construction, but only that it is recommended by the usual scientific grounds of economy and comprehensiveness of theoretical explanation.

* See Preface to Holt's *Concept of Consciousness.*

PART I

THE LOGICAL ANALYSIS OF PHYSICS

CHAPTER II

PRE-RELATIVITY PHYSICS

THE physics of Newton, considered as a deductive system, had a perfection which is absent from the physics of the present day. Science has two purposes, each of which tends to conflict with the other. On the one hand, there is a desire to know as much as possible of the facts in the region concerned; on the other hand, there is the attempt to embrace all the known facts in the smallest possible number of general laws. The law of gravitation accounted for all the facts about the motions of the planets and their satellites which were known in Newton's day; at the time, it exhibited the ideal of science. But facts and theories seem destined to conflict sooner or later. When this happens, there is a tendency either to deny the facts or to despair of theory. Thanks to Einstein, the minute facts which have been found incompatible with the natural philosophy of Newton have been fitted into a new natural philosophy; but there is not yet the complete theoretical harmony that existed while Newton was undisputed.

It is necessary to say something about the Newtonian system, since everything subsequent has arisen as an amendment to it, not as a fresh start. Most of the fundamental concepts of this system are due to Galileo, but the complete structure appears first in Newton's *Principia*. The theory is simple and mathematical; indeed, one of its main differences from modern theories is its belief (perhaps traceable to Greek geometry) that Nature is convenient for the mathematician, and requires little manipulation before his concepts become applicable.

The Newtonian system, stated with schematic simplicity, as, *e.g.*, by Boscovitch, is as follows. There is an absolute space, composed of points, and an absolute time, composed of instants; there are particles of matter, each of which persists through all time and occupies a point at each instant. Each particle exerts forces on other particles, the effect of which is to produce accelerations. Each particle is associated with a certain quantity, its " mass," which is inversely proportional to the acceleration produced in the particle by a given force. The laws of physics are conceived, on the analogy of the law of gravitation, as formulæ giving the force exerted by one particle on another in a given relative situation. This system is logically faultless. It was criticized on the ground that absolute space and time were meaningless, and on the ground that action at a distance was inconceivable. This latter objection was sanctioned by Newton, who was not a strict Newtonian. But in fact neither objection had any force from a logical point of view. Kant's antinomies, and the supposed difficulties of infinity and continuity, were finally disposed of by Georg Cantor. There was no valid *a priori* reason for supposing that Nature was not such as the Newtonians averred, and their scientific successes afforded empirical, or at least pragmatic, arguments in their favour. It is no wonder, therefore, that, throughout the eighteenth century, the system of ideas which had led to the law of gravitation dominated all scientific thought.

Before physics itself had made any breaches in this edifice, there were, however, certain objections of an epistemological order. It will be worth while to consider these, since it is urged that the theory of relativity is not open to them, though I believe this claim to be only partially justified.

The most formidable and persistent attack was upon absolute space and time. This attack was initiated by Leibniz in the lifetime of Newton, especially in his controversy with Clarke, who represented Newton. In time, most physicists came to disbelieve in absolute space and time, while retaining the Newtonian technique, which assumed their existence. In Clerk Maxwell's *Matter and Motion*, absolute motion is asserted in one passage and denied in another, with hardly any attempt to reconcile these two opinions. But at the end of the nineteenth century the prevalent view was certainly that of Mach, who vigorously denied absolute space and time. Although this denial has now been proved to be right, I cannot think that before Einstein and Minkowski it had any conclusive arguments in its favour. In spite of the fact that the whole question is now ancient history, it may be instructive to consider the arguments briefly.

The important reasons for rejecting absolute space and time were two. First, that everything we can observe has to do only with the relative positions of bodies and events; secondly, that points and instants are an unnecessary hypothesis, and are therefore to be rejected in accordance with the principle of economy, which is the same thing as Occam's razor. It appears to me that the first of these arguments has no force, while the second was false until the advent of the theory of relativity. My reasons are as follows:

That we can only observe relative positions is, of course, true; but science assumes many things that cannot be observed, for the sake of simplicity and continuity in causal laws. Leibniz assumed that there are infinitesimals, although everything that we can observe exceeds a certain minimum size. We all think that the earth has an inside, and the moon a side which we cannot see. But, it will be said, these things are like what we observe, and circumstances can be imagined under which we should observe them, whereas absolute space and time are different in kind from anything directly known, and could not be directly known in any conceivable conditions. Unfortunately, however, this applies equally to physical bodies. The relative positions which we see are relative positions of parts of the visual field; but the things in the visual field are not bodies as conceived in traditional physics, which is dominated by the Cartesian dualism of mind and matter, and places the visual field in the former. This argument is not valid as against Mach, who argued that our sensations are actually part of the physical world, and thus inaugurated the movement towards neutral monism, which denies the ultimate validity of the mind-matter dualism. But it is valid as against all those for whom matter is a sort of *Ding-an-sich*, essentially different from anything that enters into our experience. For them, it should be as illegitimate to infer matter from our perceptions as to infer absolute space and time. The one, like the other, is part of our naive beliefs, as is shown by the Copernican controversy, which would have been impossible for men who rejected absolute space and time. And the remoteness from our perceptions is as much a discovery due to reflection in the one case as in the other.

It is impossible to lay down a hard-and-fast rule that we can never validly infer something radically different from what we observe—unless, indeed, we take up the position that nothing unobserved can ever be validly inferred. This view, which is advocated by Wittgenstein in his *Tractatus Logico-Philosophicus*, has much in its favour, from the standpoint of a strict logic; but it puts an end to physics, and therefore to the problem with which this work is concerned. I shall accordingly assume that scientific inference, conducted with due care, may be valid, provided it is recognized as giving only probability, not certainty. Given this assumption, I see no possible ground for rejecting an inference to absolute space and time, if the facts seem to call for it. It may be admitted that it is better, if possible, to avoid inferring anything *very* different from what we know to exist. Such a principle will have to be based on grounds of probability. It may be said that all inferences to something unobserved are only probable, and that their probability depends, in part, upon the *a priori* probability of the hypothesis; this may be supposed greater when we infer something similar to what we know than when we infer something dissimilar. But it seems questionable whether there is much force in this argument. Everything that we perceive directly is subject to certain conditions, more especially physiological conditions; it would seem *a priori* probable that where these conditions are absent things would be different from anything that we can experience. If we suppose—as we well may—that what we experience has certain characteristics connected with our experiencing, there can be no *a priori* objection to the hypothesis that some of the things we do not experience are lacking in some characteristics which are universal in our experience. The inference to absolute space and time must, therefore, be treated as on a level with any other inductive inference.

The second argument against absolute space and time—namely, that they are unnecessary hypotheses—has turned out to be valid; but it is only in quite recent times that Newton's argument to the contrary has been refuted. The argument, as everyone knows, was concerned with absolute rotation. It is urged that, for "absolute rotation," we may substitute "rotation relatively to the fixed stars." This is formally correct, but the influence attributed to the fixed stars savours of astrology, and is scientifically incredible. Apart from this special argument, the whole of the Newtonian technique is based upon the assumption that there is such a quantity as absolute acceleration; without this, the system collapses. That is one reason why the law of gravitation cannot enter unchanged into the general theory of relativity. There are, of course, two distinct elements in the theory of relativity: one of them—the merging of space and time into space-time—is wholly new, while the other—the substitution of relative for absolute motion—has been attempted ever since the time of Leibniz. But this older problem could not be solved by itself, because of the necessity for absolute acceleration in Newtonian dynamics. Only the method of tensors, and the new law of gravitation obtained in accordance with this method, have made it possible to answer Newton's arguments for absolute space and time. While, therefore, the contention that these are unnecessary would always have been a valid ground for rejecting them if it had been known to be true, it is only now that we can be confident of its correctness, since it is only now that we possess a mathematical technique which is in accordance with it.

Somewhat similar considerations apply to action at a distance, which was also considered incredible by Newton's critics, from Leibniz onwards, and even by Newton himself. There is one theory, which may well be true, according to which action at a distance is self-contradictory: this is the theory which derives spatio-temporal separation from causal separation. I shall say no more about this possibility at present, since it was not suggested by any of the opponents of action at a distance, all of whom considered spatial and temporal relations totally distinct from causal relations. From their point of view, therefore, the objection to action at a distance seems to have been little more than a prejudice. The source of the prejudice was, I think, twofold: first, that the notion of "force," which was the dynamical form of "cause," was derived from the sensations of pushing and pulling; secondly, that people falsely supposed themselves in contact with things when they pushed and pulled them, or were pushed and pulled by them. I do not mean that such crude notions would have been explicitly defended, but that they dominated the imaginative picture of the physical world, and made Newtonian dynamics seem what is absurdly called "intelligible." Apart from such mistakes, it should have been regarded as a purely empirical question whether there is action at a distance or not. It was in fact so regarded throughout the latter half or three-quarters of the eighteenth century, and it was generally held that the empirical arguments in favour of action at a distance were overwhelming.

Not wholly unconnected with the question of action at a distance was the question of the rôle of " force " in dynamics. In Newton, " force " plays a great part, and there seems no doubt that he regarded it as a *vera causa*. If there was action at a distance, the use of the words " central forces " seemed to make it somehow more " intelligible." But gradually it was increasingly realized that " force " is merely a connecting link between configurations and accelerations; that, in fact, causal laws of the sort leading to differential equations are what we need, and that " force " is by no means necessary for the enunciation of such laws. Kirchoff and Mach developed a mechanics which dispensed with " force," and Hertz perfected their views in a treatise* comparable to Euclid from the point of view of logical beauty, leading to the result that there is only one law of motion, to the effect that, in a certain defined sense, every particle describes a geodesic. Although the whole of this development involved no essential departure from Newton, it paved the way for relativity dynamics, and provided much of the necessary mathematical apparatus, particularly in the use of the principle of least action.

The first physical theory to be developed on lines definitely different from those of Newtonian astronomy was the undulatory theory of light. Not that there was anything to contradict Newton, but that the framework of ideas was different. Transmission through a medium had been made fashionable by Descartes, and unfashionable by the Newtonians; in the case of the transmission of light it was found necessary to revert to the older point of view. Moreover, the æther was never so comfortably material as " gross " matter. It could vibrate, but it did not seem to consist of little bits

* *Prinzipien der Mechanik.*

each with its own individuality, or to be subject to any discoverable molar motions. No one knew whether it was a jelly or a gas. Its properties could not be inferred from those of billiard balls, but were merely those demanded by its functions. In fact, like a painfully good boy, it only did what it was told, and might therefore be expected to die young.

A more serious change was introduced by Faraday and Maxwell. Light had never been treated on the analogy of gravitation, but electricity appeared to consist of central forces varying inversely as the square of the distance, and was therefore confidently fitted into the Newtonian scheme. Faraday experimentally and Maxwell theoretically displayed the inadequacy of this view; Maxwell, moreover, demonstrated the identity of light and electromagnetism. The æther required for the two kinds of phenomena was therefore the same, which gave it a much better claim to be supposed to exist. Maxwell's proof, it is true, was not conclusive, but it was made so by Hertz when he produced electromagnetic waves artificially and studied their properties experimentally. It thus became clear that Maxwell's equations, which contained practically the whole of his system, must take their place beside the law of gravitation as affording the mathematical formula for a vast range of phenomena. The concepts required for these equations were, at first, not definitely contradictory to the Newtonian dynamics; but by the help of subsequent experimental results contradictions emerged which were only removed by the theory of relativity. Of this, however, we shall speak in a later chapter.

Another breach in the orthodox system, of which the importance has only become fully manifest since the publication of the general theory of relativity, was the invention of non-Euclidean geometry. In the work of Lobatchevsky and Bolyai, although the philosophical challenge to Euclid was already complete, and the consequent argument against Kant's transcendental æsthetic very powerful, there were not yet, at least obviously, the far-reaching physical implications of Riemann's inaugural dissertation "Ueber die Hypothesen, welche der Geometrie zu Grunde liegen." A few words on this topic are unavoidable at this stage, although the full discussion will come later.

One broad result of non-Euclidean geometry, even in its earliest form, was that the geometry of actual space is, at least in part, an empirical study, not a branch of pure mathematics. It may be said that empiricists, such as J. S. Mill, always based geometry upon empirical observation. But they did the same with arithmetic, in which they were certainly mistaken. No one before the non-Euclideans perceived that arithmetic and geometry stand on a quite different footing, the former being continuous with pure logic and independent of experience, the latter being continuous with physics and dependent upon physical data. Geometry can, it is true, be still studied as a branch of pure mathematics, but it is then hypothetical, and cannot claim that its initial hypotheses (which replace the axioms) are true in fact, since this is a question outside the scope of pure mathematics. The geometry which is required by the engineer or the astronomer is not a branch of pure mathematics, but a branch of physics. Indeed, in the hands of Einstein geometry has become identical with the whole of the general part of theoretical physics: the two are united in the general theory of relativity.

Riemann, who was logically the immediate predecessor of Einstein, brought in a new idea of which the importance was not perceived for half a century. He considered that geometry ought to start from the infinitesimal, and depend upon integration for statements about finite lengths, areas, or volumes. This requires, *inter alia*, the replacement of the straight line by the geodesic: the latter has a definition depending upon infinitesimal distances, while the former has not. The traditional view was that, while the length of a curve could, in general, only be defined by integration, the length of the straight line between two points could be defined as a whole, not as the limit of a sum of little bits. Riemann's view was that a straight line does not differ from a curve in this respect. Moreover, measurement, being performed by means of bodies, is a physical operation, and its results depend for their interpretation upon the laws of physics. This point of view has turned out to be of very great importance. Its scope has been extended by the theory of relativity, but in essence it is to be found in Riemann's dissertation.

Riemann's work, as well as that of Faraday and Maxwell, belongs, like the theory of relativity, to the development of the view of the physical world as a continuous medium, which has, from the earliest times, contested the mastery with the atomic view. Just as Newton caused absolute space and time to be embedded in the technique of dynamics, so Pythagoras caused spatial atomism to be embedded in the technique of geometry. Ever since Greek times, those who did not believe in the reality of " points " were faced with the difficulty that a geometry based on points works, while no other way of starting geometry was known. This difficulty, as Dr Whitehead has shown, exists no longer. It is now possible, as we shall see at a later stage, to interpret geometry and physics with material all of which is of a finite size—it is even possible to demand that none of the material shall be smaller than an assigned finite size. The fact that this hypothesis can be reconciled with mathematical continuity is a novel discovery of considerable importance; until recently, atomism and continuity appeared incompatible. There are, however, forms of atomism which have not hitherto been found easy to reconcile with continuity; and, as it happens, there is powerful experimental evidence in their favour. Just at the moment when Maxwell, supplemented by Hertz, appeared to have reduced everything to continuity, the new evidence for an atomic view of Nature began to accumulate. There is still an unreconciled conflict, one set of facts pointing in one direction, and another in another; but it is legitimate to hope that the conflict will be resolved before long. Modern atomism, however, demands a new chapter.

CHAPTER III

ELECTRONS AND PROTONS

PHYSICS, at the present time, is divisible into two parts, the one dealing with the propagation of energy in matter or in regions where there is no matter, the other with the interchanges of energy between these regions and matter. The former is found to require continuity, the latter discontinuity. But before considering this apparent conflict, it will be advisable to deal in outline with the discontinuous characteristics of matter and energy as they appear in the theory of quanta and in the structure of atoms. It is necessary, however, for philosophical purposes, to deal only with the most general aspects of modern theories, since the subject is developing rapidly, and any statement runs a risk of being out of date before it can be printed. The topics considered in this chapter and the next have been treated in an entirely new way by the theory initiated by Heisenberg in 1925. I shall, however, postpone the consideration of this theory until after that of the Rutherford-Bohr atom and the theory of quanta connected with it.

It appears that both matter and electricity are concentrated exclusively in certain finite units, called electrons and protons. It is possible that the helium nucleus may be a third independent unit, but this seems improbable.* The net positive charge of a helium nucleus is double that of a proton, and its mass is slightly less than four times that of a proton. These facts are explicable (including the slight deficiency of mass) if the

* Professors F. Paneth and K. Peters claim to have transformed hydrogen into helium. If this claim is substantiated, it disposes definitively of the possibility that the helium nucleus is an independent unit. See *Nature*, October 9, 1926, p. 526.

helium nucleus consists of four protons and two electrons; otherwise, they seem an almost incredible coincidence. We may therefore assume that electrons and protons are the sole constituents of matter; if it should turn out that the helium nucleus must be added, that would make little difference to the philosophical analysis of matter, which is our task in this volume.

Protons all have the same mass and the same amount of positive electricity. Electrons all have the same mass, about $\frac{1}{1835}$ of the mass of a proton. The amount of negative electricity on an electron is always the same, and is such as to balance exactly the amount on a proton, so that one electron and one proton together constitute an electrically neutral system. An atom consists, when unelectrified, of a nucleus surrounded by planetary electrons: the number of these electrons is the atomic number of the element concerned. The nucleus consists of protons and electrons: the number of the former is the atomic weight of the element, the number of the latter is such as to make the whole electrically neutral—*i.e.* it is the difference between the number of protons in the nucleus and the number of planetary electrons. Every item in this complicated structure is supposed, at normal times, to be engaged in motions which result, on Newtonian principles (modified slightly by relativity considerations), from the attractions between electrons and protons and the repulsions between protons and protons as well as between electrons and electrons. But of all the motions which should be possible on the analogy of the solar system, it is held that only an infinitesimal proportion are in fact possible; this depends upon the theory of quanta, in ways which we shall consider later.

The calculation of the orbits of planetary electrons, on Newtonian principles, is only possible in the two simplest cases: that of hydrogen, which consists (when unelectrified) of one proton and one electron; and that of positively electrified helium, which has lost one, but not both, of its planetary electrons. In these two cases the mathematical theory is practically complete. In all other cases which actually occur, although the mathematics required is of a sort which has been investigated ever since the time of Newton, it is impossible to obtain exact solutions, or even good approximations. The case is still worse as regards nuclei. The nucleus of hydrogen is a single proton, but that of the next element, helium, is held to consist of four protons and two electrons. The combination must be extraordinarily stable, both because no known process disintegrates the helium nucleus, and because of the loss of mass involved. (If the mass of the helium atom is taken as 4, that of a hydrogen atom is not 1, but 1·008.) This latter argument depends upon considerations connected with relativity, and must therefore be discussed at a later stage. Various suggestions have been made as to the way in which the protons and electrons are arranged in the helium nucleus, but none, so far, has yielded the necessary stability. What we may call the geometry of nuclei is therefore still unknown. It may be that, at the very small distances involved, the law of force is not the inverse square, although this law is found perfectly satisfactory in dealing with the motions of the planetary electron in the two cases in which the mathematics is feasible. This, however, is merely a speculation; for the present we must be content with ignorance as regards the arrangement of protons and electrons in nuclei other than that of hydrogen (which contains no electron in the nucleus).

So long as an atom remains in a state of steady motion, it gives no evidence of its existence to the outside world. A material system displays its existence to outsiders by radiating or absorbing energy, and in no other way; and an atom does not absorb or part with energy except when it undergoes sudden revolutionary changes of the sort considered by the theory of quanta. This is of importance from our point of view, since it shows that no empirical evidence can decide between two theories of the atom which yield the same result as regards the interchanges of energy between the atom and the surrounding medium. It may be that the whole Rutherford-Bohr theory is too concrete and pictorial; the analogy with the solar system may be much less close than it is represented as being. A theory which accounts for all the known facts is not thereby shown to be true: this would require a proof that no other theory would do the same. Such a proof is very seldom possible; certainly it is not possible in the case of the structure of the atom. What may be taken as firm ground is the numerical part of the theory. Certain quantities, and certain whole numbers, are clearly involved; but it would be rash to say that such and such an interpretation of these quantities and whole numbers is the only one possible. It is proper and right to use a pictorial theory as a help in investigation; but what can count as definite knowledge is something much more abstract. And it is quite possible that the truth does not lend itself to pictorial statement, but only to expression in mathematical formulæ. This, as we shall see, is the view taken by what we may call the Heisenberg theory.

It may be worth while to linger a moment over this question of the nature of our real knowledge concerning atoms. In the last analysis, all our knowledge of matter is derived from perceptions, which are themselves causally dependent upon effects on our body. In sight, for example, we depend upon light-waves which impinge upon the eye. Given the waves, we shall have the visual perception, assuming no defect in the eye. Therefore nothing in visual perception alone can enable us to distinguish between two theories which give the same result as regards the light-waves which reach human eyes. This, as stated, seems to introduce psychological considerations. But we may put the matter in a way that makes its physical significance clearer. Consider an oval surface, which is liable to continuous motion and change of shape, but persists throughout time; and let us suppose that no human being has ever been inside this surface. In illustration, we might take a sphere surrounding the sun, or a little box surrounding an electron which never forms part of a human body. Energy will cross this surface, sometimes inward, sometimes outward. Two views which lead to the same results as to the flow of energy across the boundary are empirically indistinguishable, since everything that we know independently of physical theory lies outside the surface. We may enlarge our oval surface until its " inside " consists of everything outside the body of the physicist concerned—to wit, ourselves. What we hear, and what we read in books, comes to us entirely through a flow of energy across the boundary of our body. It may well be maintained that our direct knowledge is less than this statement would imply, but it is certainly not greater. Two universes which give the same results for the flow of energy across the boundary of A's body will be totally indistinguishable for A.

My object in bringing up these considerations is partly to give a new turn to the argument about solipsism. As a rule, solipsism is taken as a form of idealism—namely, the view that nothing exists except my mind and my mental events. I think, however, that it would be just as rational, or just as irrational, to say that nothing exists outside my body, or that nothing exists outside a certain closed surface which includes my body. Neither of these is the general form of the argument. The general form is that first given above—namely, that, given any region not containing myself, two physical theories which give the same boundary conditions all over this region are empirically indistinguishable. Electrons and protons, in particular, are only known by their effects elsewhere, and so long as these effects are unchanged we may alter our views of electrons and protons as much as we please without making a difference in anything verifiable. The question of the validity of the inference to things outside ourselves is logically quite distinct from the question whether the stuff of the world is mental, material, or neutral. I might be a solipsist, and hold at the same time that I am my body; I might, conversely, allow inferences to things other than myself, but maintain that these things were minds or mental events. In physics, the question is not that of solipsism, but the much more definite question: Given the physical conditions at the bounding surface of some volume, without any direct knowledge of the interior, how much can we legitimately infer as to what happens in the interior? Is there good ground for supposing that we can infer as much as physicists usually assume? Or can we perhaps infer much less than is generally supposed? I do not propose as yet to attempt an answer to this question; I have raised it at this stage in order to suggest a doubt as to the completeness of our knowledge concerning the structure of the atom.

CHAPTER IV

THE THEORY OF QUANTA

THE atomicity of matter is a hypothesis as old as the Greeks, and in no way repugnant to our mental habits. The theory that matter is composed of electrons and protons is beautiful through its successful simplicity, but is not difficult to imagine or believe. It is otherwise with the form of atomicity introduced by the theory of quanta. This might possibly not have surprised Pythagoras, but it would most certainly have astonished every later man of science, as it has astonished those of our own day. It is necessary to understand the general principles of the theory before attempting a modern philosophy of matter; but unfortunately there are still unsolved physical problems connected with it, which make it improbable that a satisfactory philosophy of the subject can yet be constructed. Nevertheless, we must do what we can.

As everyone knows, the quantum was first introduced by Planck in 1900 in his study of black-body radiation. Planck showed that, when we consider the vibrations which constitute the heat in a body, these are not distributed among all possible values according to the usual law of frequency which governs chance distributions, but on the contrary are tied down by a certain law. If ϵ is the energy of a vibration, and ν its frequency, then there is a certain constant h,* known as Planck's constant, such that ϵ/ν is h, or $2h$, or $3h$, or some other small integral multiple of h. Vibrations with other amounts of energy do n. t occur. No reason is known for their non-occurrence, which remains so far of the nature of a brute

* The numerical value of h is $6{\cdot}55 \times 10^{-27}$ erg secs., and its dimensions are those of " action "—*i.e.*, energy $\times$ time.

fact. At first, it was an isolated fact. But now Planck's constant has been found to be involved in various other kinds of phenomena; in fact, wherever observation is sufficiently minute to make it possible to discover whether it is involved or not.

A second field for the quantum theory was found in the photo-electric effect. This effect is described as follows by Jeans:*

"The general features of the phenomenon are well known. For some time it has been known that the incidence of high-frequency light on to the surface of a negatively charged conductor tended to precipitate a discharge, while Hertz showed that the incidence of the light on an uncharged conductor resulted in its acquiring a positive charge. These phenomena have been shown quite conclusively to depend on the emission of electrons from the surface of the metal, the electrons being set free in some way by the incidence of the light.

"In any particular experiment, the velocities with which individual electrons leave the metal have all values from zero up to a certain maximum velocity v, which depends on the conditions of the particular experiment. No electron is found to leave the metal with a velocity greater than this maximum v. It seems probable that in any one experiment all the electrons are initially shot off with the same velocity v, but that those which come from a small distance below the surface lose part of their velocity in fighting their way out to the surface.

"Leaving out of account such disturbing influences as films of impurities on the metallic surface, it appears to be a general law that the maximum velocity v depends only on the nature of the metal and on the *frequency* of the incident light. It does not depend on the intensity of the light, and within the range of temperature within which experiments are possible it does not depend on the temperature of the metal. . . . For a given metal this maximum velocity increases regularly as the frequency of the light is increased, but there is a certain frequency below which no emission takes place at all."

* *Report on Radiation and the Quantum Theory*, Physical Society of London, 1914, p. 58.

The explanation of this phenomenon in terms of the quantum was first given by Einstein* in 1905. When light of frequency ν falls on the conductor, it is found that the amount of energy absorbed by an electron which the light separates from its atom is about five-sixths of $h\nu$, where ν is Planck's constant. It may be supposed that the other one-sixth is absorbed by the atom, so that atom and electron together absorb exactly one quantum h. When the light is of such low frequency that $h\nu$ is not enough to liberate an electron, the photo-electric effect does not take place. Explanations not involving the quantum have been attempted, but none seem able to account for the data.

Another field in which the quantum hypothesis has been found necessary is the specific heat of solids at low temperatures. According to previous theories, the specific heat (at constant volume) multiplied by the atomic weight ought to have the constant value 5·95. In fact, this is found to be very approximately correct for high temperatures, but for low temperatures there is a falling off which increases as the temperature falls. The explanation of this fact offered by Debye is closely analogous to Planck's explanation of the facts of black-body radiation; and as in that case, it seems definitely impossible to obtain a satisfactory theory without invoking the quantum.†

The most interesting application of quantum theory is Bohr's explanation of the line spectra of elements. It had been found empirically that the lines in the hydrogen spectrum which were known had frequencies obtained from the difference of two "terms," according to the formula:

$$\nu = R\left(\frac{1}{n^2} - \frac{1}{k^2}\right) \quad \ldots\ldots\ldots\ldots\ldots\ldots (1),$$

where ν is the frequency, R is "Rydberg's constant," n and k

* *Annalen der Physik*, vol. xvii., p. 146.
† See Jeans, *loc. cit.*, chap. vi.

are small integers, and $\frac{R}{n^2}, \frac{R}{k^2}$ are what are called "terms." After the formula had been discovered, new lines agreeing with it were sought and found. Certain lines formerly attributed to hydrogen, and not agreeing with the above formula, were attributed by Bohr to ionized helium; they are given by the formulæ:

$$\nu = 4R\left(\frac{1}{3^2} - \frac{1}{k^2}\right)$$

$$\nu = 4R\left(\frac{1}{4^2} - \frac{1}{k^2}\right).$$

Bohr's theoretical grounds for attributing these lines to helium were afterwards confirmed experimentally by Fowler. It will be seen that they fit into the formula (1) when $4R$ is substituted for R, a fact which Bohr's theory explains, as well as the more delicate fact that, to make the formula exact, we have to substitute, not exactly $4R$, but a slightly smaller quantity.

The form of the equation (1) suggested to Bohr that a line of the hydrogen spectrum is not to be regarded as something which the atom emits when it is in a state of periodic vibration, but as produced by a change from a state connected with one integer to a state connected with another. This would be explained if the orbit of the electron were not just any orbit possible on Newtonian principles, but only an orbit connected with an integral "quantum number"—*i.e.* with a multiple of h.

The way in which Bohr achieved a theory on these lines is as follows. He supposed that the electron can only revolve round the nucleus in certain circles, these being such that, if p is the moment of momentum in any orbit, we shall have:

$$2\pi p = nh \quad \dots\dots\dots\dots\dots\dots (2),$$

where h is, as always, Planck's constant, and n is a small whole number. (In theory n might be any whole number, but in practice it is never found to be much larger than 30, and that only in certain very tenuous nebulæ.) The reason why the quantum principle assumes just this form will be explained presently.

Now if m is the mass of the electron, a the radius of its orbit, and ω its angular velocity, we have:

$$p = ma^2\omega.$$

Hence
$$2\pi ma^2\omega = nh \quad \ldots\ldots\ldots\ldots\ldots (3).$$

But, on grounds of the usual theory, since the radial acceleration of the electron is $a\omega^2$ and the force attracting it to the nucleus is e^2/a^2, we have:

$$ma\omega^2 = e^2/a^2.$$

I.e.
$$ma^3\omega^2 = e^2 \quad \ldots\ldots\ldots\ldots\ldots\ldots (4).$$

From equations (3) and (4) we obtain:

$$a = \frac{n^2h^2}{4\pi^2me^2}, \quad \omega = \frac{8\pi^3me^4}{n^3h^3} \quad \ldots\ldots\ldots\ldots (5).$$

The possible orbits for the electron are obtained by putting $n = 1, 2, 3, 4, \ldots$ in the above formulæ for a. Thus the smallest possible orbit is:

$$a_1 = \frac{h^2}{4\pi me^2} \quad \ldots\ldots\ldots\ldots\ldots\ldots (6);$$

and the other possible orbits are $4a_1$, $9a_1$, $16a_1$, etc.

For the energy in an orbit of radius n^2a_1 we have, since the potential energy is double the kinetic energy with its sign changed:*

$$W = -\tfrac{1}{2}ma^2\omega^2 = -\frac{2\pi^2me^4}{n^2h^2}$$

in virtue of (5). Thus when the electron falls from an orbit whose radius is k^2a_1 to one whose radius is n^2a_1 $(k > n)$, there is a loss of energy:

$$\frac{2\pi^2me^4}{h^2}\left(\frac{1}{n^2} - \frac{1}{k^2}\right).$$

* See Sommerfeld, *Atomic Structure and Spectral Lines*, pp. 547 ff.

It is assumed that this energy is radiated out in a light-wave whose energy is one quantum of energy $h\nu$, where ν is its frequency. Hence we obtain the frequency of the emitted light by the equation:

$$h\nu = \frac{2\pi^2 me^4}{h^2}\left(\frac{1}{n^2}-\frac{1}{k^2}\right);$$

i.e.

$$\nu = \frac{2\pi^2 me^4}{h^3}\left(\frac{1}{n^2}-\frac{1}{k^2}\right).$$

This agrees exactly with the observed lines if [see equation (1)]:

$$R = \frac{2\pi^2 me^4}{h^3},$$

where R is Rydberg's constant. On inserting numerical values, it is found that this equation is verified. This striking success was, from the first, a powerful argument in favour of Bohr's theory.

Bohr's theory has been generalized by Wilson* and Sommerfeld so as to allow also elliptic orbits: these have two quantum numbers, one corresponding, as before, to angular momentum or the moment of momentum (which is constant, by Kepler's second law), the other depending upon the eccentricity. Only certain eccentricities are possible; in fact, the ratio of the minor to the major axis is always rational, and has as its denominator the quantum number corresponding to the moment of momentum. In order to explain the Zeeman effect (which arises in a magnetic field) we used a third quantum number, corresponding to the angle between the plane of the magnetic field and the plane of the electron's orbit. In all cases, however, there is a general principle, which must now be explained. This will show, also, why, in Bohr's theory, the quantum equation (2) takes the form it does.†

* W. Wilson, *The Quantum Theory of Radiation and Line Spectra* Phil. Mag., June, 1915.

† What follows is taken from Note 7 (pp. 555 ff.) in Sommerfeld's *Atomic Structure and Spectral Lines*, translated from the third German edition by Henry L. Brose, M.A., 1923. See also Note 4 (pp. 541 ff.).

The first thing to observe is that the quantum principle is really concerned with atoms of *action*, not of *energy*: action is energy multiplied by time. Suppose now that we have a system depending upon several co-ordinates, and periodic in respect of each. It is not necessary to suppose that each co-ordinate has the same period: it is only necessary to suppose that the system is "conditionally periodic"—*i.e.* that each co-ordinate separately is periodic. We must further assume that our co-ordinates are so chosen as to allow "separation of variables" (as to which, see Sommerfeld, *op. cit.*, pp. 559-60). We then define the "momentum" (in a generalized sense) associated with the co-ordinate q_k as the partial differential of the kinetic energy with respect to $\dot{q}_k$—*i.e.* calling the generalized momentum p_k, we put:

$$p_k = \frac{\partial E_{kin}}{\partial \dot{q}_k},$$

where E_{kin} is the kinetic energy. The quantum condition is to apply to the integral of p_k over a complete period of q_k—*i.e.* we are to have:

$$\int p_k dq_k = n_k h,$$

where the integration is taken through one complete period of q_k. Here n_k will be the quantum number associated with the co-ordinate q_k. The above is a general formula of which all known cases of quantum phenomena are special cases. This is its sole justification.

The above principle is exceedingly complicated—more so, even, than it appears in our summary account, which has omitted various difficulties. It is possible that its complication may be due to the fact that quantum dynamics has had to force its way through the obstacles which the classical system put in its way; it is possible also that quantum phenomena may turn out to be deducible from classical principles. But before pursuing this line of thought, it may be well to say a few words about the developments of Bohr's theory by Sommerfeld and others.

In its original form, in which circular orbits were assumed, Bohr's theory accounted for the main facts concerning the line spectra of hydrogen and ionized helium. But there were a number of more delicate facts which required the hypothesis of elliptic orbits: with this hypothesis, together with some niceties derived from relativity, the most minute agreement has been obtained between theory and observation. But perhaps this great success has made people think that more was proved than really was proved. The great advantage obtained from admitting elliptic orbits is that they provide a second quantum number. In the emission of light by atoms, what we have is essentially as follows. The atom is capable of various states, characterized by whole numbers (the quantum numbers). There may be more or fewer quantum numbers, according to the degrees of freedom of the system. The loss or gain of energy when an atom passes from a state characterized by one set of values of the quantum numbers to a state characterized by another set is known. When energy is lost (without the loss of an electron or of any part of the nucleus of the atom), it passes out as a light-wave, whose energy is equal to what the atom has lost, and whose energy multiplied by the time of one vibration is h. Energy is what is conserved, but action is what is quantized.

Let us revert, in illustration, to the circular orbits of Bohr's original theory, which remain possible, though not universal, in the newer theory. If we call E_{min} the kinetic energy when the electron is in the smallest possible orbit, the kinetic energy in the nth orbit is $\frac{E_{min}}{n^2}$. (The measure of the total energy is the kinetic energy with its sign changed.) We do not know what determines the electron to jump from one orbit to another; on this point, our knowledge is merely statistical.

We know, of course, that when the atom is not in a position to absorb energy the electron can only jump from a larger to a smaller orbit, while the converse jump occurs when the atom absorbs energy from incident light. We know also, from the comparative intensities of different lines in the spectrum, the comparative frequencies of different possible jumps, and on this subject a theory exists. But we do not know in the least why, of a number of atoms whose electrons are not in minimum orbits, some jump at one time and some at another, just as we do not know why some atoms of radio-active substances break down while others do not. Nature seems to be full of revolutionary occurrences as to which we can say that, *if* they take place, they will be of one of several possible kinds, but we cannot say that they will take place at all, or, if they will, at what time. So far as quantum theory can say at present, atoms might as well be possessed of free will, limited, however, to one of several possible choices.*

However this may be, it is clear that what we know is the changes of energy when an atom emits light, and we know that in the case of hydrogen or ionized helium these changes are measured by $\frac{1}{n^2} - \frac{1}{k^2}$. It seems almost unavoidable to infer that the previous state of the atom was characterized by the integer k and the later one by the integer n. But to assume orbits and so on, though proper as a help to the imagination, is hardly sufficiently justified by the analogy of large-scale processes, since the quantum principle itself shows the danger of relying upon this analogy. In large-scale occurrences there is nothing to suggest the quantum, and perhaps other familiar features of such occurrences may result merely from statistical averaging.

* This, however, is probably a temporary state of affairs. Certain Pasons for quantum transitions are already known. See J. Franck and P. Jordan, *Anregung von Quantensprüngen durch Stösse*, Berlin, 1926; also P. Jordan, *Kausalität und Statistik in der modernen Physik*, *Naturwissenschaften*, Feb. 4, 1927.

It may be worth while to consider briefly the elliptical orbits which are possible.* This will also illustrate the application of the quantum principle to systems with more than one co-ordinate.

Taking polar co-ordinates, the kinetic energy is:

$$\tfrac{1}{2}m(\dot{r}^2+r^2\dot{\theta}^2).$$

The two generalized momenta are therefore:

$$p_\theta = mr^2\dot{\theta},\ p_r = m\dot{r}.$$

We have thus two quantum conditions:

$$\int_0^{2\pi} 2mr^2\dot{\theta}\,d\theta = nh$$

and

$$\int_{\theta=0}^{\theta=2\pi} m\dot{r}\,dr = n'h.$$

By Kepler's second law, $mr^2\dot{\theta}$ is constant; call it p. Thus:

$$2\pi p = nh.$$

The other integration is more troublesome, but we arrive at the result that, if a and b are the major and minor axes of the ellipse,

$$\frac{a-b}{a}=\frac{n'}{n}.$$

A little further calculation leads to the result that the energy in the orbit which has the quantum numbers n, n' is:

$$-\frac{2\pi^2me^4}{h^2}\cdot\frac{1}{(n+n')^2}.$$

This is exactly the same as in the case of circular orbits, except that $n+n'$ replaces n. If this were all, the line spectrum of hydrogen would be exactly the same whether elliptic orbits occurred or not, and there would be no empirical means of deciding the question.

However, by introducing considerations derived from the special theory of relativity we are able to distinguish between the results to be expected from circular and elliptic orbits

* See Sommerfeld, *op. cit.*, pp. 232 ff.

respectively, and to show that the latter must occur to account for observed facts. The crucial point is the variation of mass with velocity: the faster a body is moving, the greater is its mass. Therefore in an elliptic orbit the electron will have a greater mass at the perihelion than at the aphelion. From this it is found to follow that an elliptic orbit will not be accurately elliptic, but that the perihelion will advance slightly with each revolution.* That is to say, taking polar co-ordinates r, θ, the co-ordinate θ increases by slightly more than 2π between one minimum of r and the next. The system is thus "conditionally periodic"—*i.e.* each separate co-ordinate changes periodically, but the periods of the two do not coincide. The result† is that the equation $\frac{a-b}{a}=\frac{n'}{n}$ is replaced by:

$$\frac{a-b}{a}=\frac{n'}{n\gamma},$$

where

$$\gamma^2=1-\frac{e^4}{c^2p^2},$$

c being the velocity of light, and p, as before, the angular momentum. It will be seen that γ is very nearly 1, because c is large.

The formula for the energy associated with the quantum numbers n, n' now becomes much more complicated; its great merit is that it accounts for the fine structure of the hydrogen line spectrum. It must be felt that this minuteness of agreement between theory and observation is very remarkable. But it is still the case that the only empirical evidence concerns differences of energy in connection with different quantum numbers, and that the theory of actual orbits, proceeding, during steady motion, according to Newtonian principles, must inevitably remain a hypothesis—a hypothesis which,

* This is not the same phenomenon as in the case of the orbit of Mercury. The latter depends upon the general theory of relativity, the fomrer upon the special theory.

† Sommerfeld, *op. cit.*, pp. 467 ff.

as we shall see, has disappeared from the latest form of the quantum theory.

The fact of the existence of the quantum is as strange as it is undeniable, unless it should turn out to be deducible from classical principles. It seems to be the case that quantum principles regulate all interchange of energy between matter and the surrounding medium. There are grave difficulties in reconciling the quantum theory with the undulatory theory of light, but we shall not consider these until a later stage. What is much to be wished is some way of formulating the quantum principle which shall be less strange and *ad hoc* than that due to Wilson and Sommerfeld. For practical purposes, it amounts to something like this: that a periodic process of frequency ν has an amount of energy which is a multiple of $h\nu$, and, conversely, if a given amount of energy is expended in starting a periodic process, it will start a process with a frequency ν such that the given amount of energy shall be a multiple of $h\nu$. When a process has a frequency ν and an energy $h\nu$, the amount of "action" during one period is h. But we cannot say: In any periodic process the amount of action in one period is h or a multiple of h. Nevertheless, some formulation analogous to this might in time turn out to be possible. As has appeared from the theory of relativity, "action" is more fundamental than energy in physical theory; it is therefore perhaps not surprising that action should be found to play an important part. But the whole theory of the interaction of matter and the surrounding medium, at present, rests upon the conservation of energy. Perhaps a theory giving more prominence to action may be possible, and may facilitate a simpler statement of the quantum principle.

In Bohr's theory and its developments, there is a lacuna and there is a difficulty. The lacuna has already been mentioned: we do not know in the least why an electron chooses one moment rather than another to jump from a larger to a smaller orbit. The difficulty is that the jump is usually regarded as sudden and discontinuous: it is suggested that if it were continuous, the experimental facts in the regions concerned would become inexplicable. Possibly this difficulty may be overcome, and it may be found that the transition from one orbit to another can be continuous. But it is as well to consider the other possibility, that the transition is really discontinuous. I have emphasized how little we really *know* about what goes on in the atom, because I wished to keep open the possibility of something quite different from what is usually supposed. Have we any good reason for thinking that space-time is continuous? Do we know that, between one orbit and the next, other orbits are *geometrically* possible? Einstein has led us to think that the neighbourhood of matter makes space non-Euclidean; might it not also make it discontinuous? It is certainly rash to assume that the minute structure of the world resembles that which is found to suit large-scale phenomena, which may be only statistical averages. These considerations may serve as an introduction to the most modern theory of quantum mechanics, to which we must now turn our attention.*

In the new theory inaugurated by Heisenberg, we no longer have the simplicity of the Rutherford-Bohr atom, in which electrons revolve about a nucleus like separate planets.

* The principal papers setting forth this theory are:

1. W. Heisenberg, *Ueber quantentheoretische Umdeutung kinematischer und mechanischer Beziehungen.* Zeitschrift für Physik, 33, pp. 879-893, 1925.
2. M. Born and P. Jordan, *Zur Quantenmechanik. Ibid.* 34, pp. 858-888, 1925.
3. M. Born, W. Heisenberg, and P. Jordan, *Zur Quantenmechanik II. Ibid.* 35, pp. 557-615, 1926.
4. P. A. M. Dirac, *The Fundamental Equations of Quantum Mechanics.* Proc. Royal Soc., Series A, vol. 109, No. A752, pp. 642-653, 1925.
5. W. Heisenberg, *Ueber quantentheoretische Kinematik und Mechanik.* Mathematische Annalen, 95, pp. 683-705, 1926.
6. W. Heisenberg, *Quantenmechanik.* Naturwissenschaften, 14 Jahrgang, Heft 45, pp. 989-994.

I shall quote these papers by the above numbers. I am much indebted in this matter to Mr R. H. Fowler, F.R.S.

Heisenberg points out that in this theory there are many quantities which are not even theoretically observable—namely, those representing processes supposed to be occurring while the atom is in a steady state. In the new theory, as Dirac says: "The variable quantities associated with a stationary state on Bohr's theory, the amplitudes and frequencies of orbital motion, have no physical meaning and are of no physical importance" (4, p. 652). Heisenberg, in first introducing his theory, pointed out that the ordinary quantum theory uses unobservable quantities, such as the position and time of revolution of an electron (1, p. 879), and that the electron ought to be represented by measurable quantities such as the frequencies of its radiation (1, p. 880). Now the observable frequencies are always differences between two "terms," each of which is represented by an integer. We thus arrive at a representation of the state of an atom by means of an infinite array of numbers—*i.e.* by a matrix. If T_n and T_m are two "terms," an observable frequency (in theory) is ν_{nm}, where:

$$\nu_{nm} = T_n - T_m.$$

It is such numbers as ν_{nm} (of which there is a doubly infinite series) that characterize the atom, so far as it is observable.

Heisenberg sets out this view as follows (5, p. 685). In the classical theory, given an electron with one degree of freedom, in harmonic oscillation, the elongation x at time t can be represented by a Fourier series:

$$x = x(n, t) = \sum_{\tau} x(n)_\tau . e^{2\pi i\nu(n).\tau.t},$$

where n is a constant and τ is the number of the harmonic. The single terms of this series, namely:

$$x(n)_\tau e^{2\pi i\nu(n)\tau t},$$

would contain the quantities which have been signalized as directly observable—namely, frequency, amplitude, and phase.

But in virtue of the fact that, in atoms, frequencies are found to be the differences of " terms," we shall have to replace the above by:

$$x(nm)e^{2\pi i\nu(nm)\tau t};$$

and the collection (not the sum) of such terms represents what was formerly the elongation x. The sum of all these terms has no longer any physical significance. Thus the atom comes to be represented by the numbers $\nu(nm)$, arranged in an infinite rectangle or " matrix."

It is possible to construct an algebra of matrices, which differs formally from ordinary algebra in only one respect, namely, that multiplication is not commutative.

A new operation is defined which, when the quantum numbers become large, approximates to differentiation. By using this operation, Hamilton's equations of motion can be preserved in a form which is applicable equally to periodic and to unperiodic motions, so that it is no longer necessary to distinguish a certain sphere of quantum phenomena, to which different laws are applied from those applied to the phenomena amenable to classical dynamics: " A distinction between 'quantized' and 'unquantized' motions loses all meaning in this theory, since in it there is no question of a quantum condition which selects certain motions from a great number of possible ones; in place of this condition appears a quantum-mechanical fundamental equation . . . which is valid for *all* possible motions, and is necessary in order to give a definite meaning to the problem of motion " (3, p. 558). The fundamental equation alluded to in the above is as follows: Let q be a Hamiltonian co-ordinate, and p the corresponding (generalized) momentum, both being matrices. It will be remembered that multiplication is not commutative for matrices; in fact, we have as the fundamental equation in question (2, p. 871):

$$pq - qp = \frac{h}{2\pi i}.1,$$

where 1 represents the matrix whose diagonal consists of 1's, and whose other terms are all zero. The above is the sole fundamental equation containing h (Planck's constant), and it is true for *all* motions.

Heisenberg does not claim that the new theory solves all difficulties. On the contrary, he says (5, p. 705):

"The theory here described must be regarded as still incomplete. The real geometrical or kinematical meaning of the fundamental assumption (5)* has not yet been made completely clear. In particular, there is a serious difficulty in the fact that the time apparently has a different rôle from the space co-ordinates, and is formally differently treated. The formal character of the time co-ordinate in the mathematical structure of the theory is made particularly evident by the fact that in the theory hitherto the question of the temporal course of a process has no immediate meaning, and that the concept of earlier and later can hardly be defined exactly. Nevertheless, we need not consider these difficulties as an objection to the theory, since the appearance of just such difficulties was to be expected from the nature of the space-time relations that hold for atomic systems."

In a more or less popular exposition (6), Heisenberg has set forth some of the consequences of his theory. Electrons and atoms, he says, do not have "the degree of immediate reality of objects of sense," but only the sort of reality which one naturally ascribes to light quanta. The troubles of the quantum theory have come, he thinks, from trying to make models of atoms and picture them as in ordinary space. If we are to retain the corpuscular theory, we can only do it by not assigning a definite point of space at each time to the electron or atom. We substitute a well-defined physical group of quantities which represent what *was* the place of the electron.

* This is the assumption, mentioned above, that an atom or electron at time t can be represented by a collection of terms of the form:

$$x(nm)e^{2\ i\nu(mn)t}.$$

They are the observable radiation quantities, each of which is associated with two " terms," so that we obtain a matrix. The distinction of inner and outer electrons in an atom becomes meaningless. " It is, moreover, in principle impossible to identify again a particular corpuscle among a series of similar corpuscles " (p. 993).

The matrix theory of the electron is too new to be amenable, as yet, to the kind of logical analysis which it is our purpose to undertake in this Part. It is clear, however, that it affects a scientific economy by substituting for the merely hypothetical steady motions of Bohr's atoms a set of quantities representing what we really know—namely, the radiations that come out of the region in which the atom is supposed to be. It is clear, also, that there is an immense logical progress in the construction of a dynamic which destroys the distinction between quantized and unquantized motions, and treats all motions by means of a uniform set of principles. And the greater abstractness of the Heisenberg atom as compared with the Bohr atom makes it logically preferable, since the pictorial elements in a physical theory are those upon which least reliance can be placed.

An apparently different quantum theory, due to de Broglie* and Schrödinger,† has been found to be formally the same as Heisinger's theory, although at first sight very different. This is described by de Broglie as " the new wave theory of matter," in which " the material point is conceived as a singularity in a wave."‡ Here, also, the radiations which we think of as coming out of the atom have more physical " reality " than the atom itself. One of the merits of the theory is that it diminishes the difficulties hitherto

* *Annales de Physique*, 3, 22, 1925.
† *Annalen der Physik*, 1926. Four papers, 79, pp. 361, 489, 734; 80, p. 437.
‡ *Nature*, Sp. 25, 1926, p. 441. See also Fowler, "Matrix and Wave Mechanics," *ib.* Feb. 12, 1927.

existing in the way of a reconciliation of the facts of interference and dispersion with the facts which led to the hypothesis of light quanta.

Meanwhile, there remains the possibility that all the quantum phenomena may be deducible from classical principles, and that the apparent discontinuities may be only a question of sharp maxima or minima. The most successful theory known to me on these lines is that of L. V. King.* He assumes that electrons rotate with a certain fixed angular velocity, the same for all; he makes a similar assumption as regards protons. Consequently there is a magnetic field which introduces conditions that are absent if electrons and protons have no spin. There will be electromagnetic radiation of frequency ν, where:

$$h\nu = \tfrac{1}{2}m_0 v^2,$$

h being Planck's constant, m_0 the invariant mass of the electron, and v its velocity. (The identity of h with Planck's constant is obtained by adjusting the hypothetical constants.) From this formula he deduces many of the phenomena upon which the quantum theory is based, and promises to deduce others in a later paper. An article by Mr R. H. Fowler ("Spinning Electrons," *Nature*, Jan. 15, 1927) discusses Mr King's theory without arriving at a verdict for or against. Presumably it will not be long before a definite answer as to the adequacy of Mr King's theory is possible. If it is adequate, the quantum theory ceases to concern the philosopher, since what remains valid in it becomes a deduction from more fundamental laws and processes which are continuous and involve no atomicity of action. For the moment, until the physicists have arrived at a decision, the philosopher must be content to investigate both hypotheses impartially.

* *Gyromagnetic Electrons and a Classical Theory of Atomic Structure and Radiation.* By Louis Vessot King, F.R.S., Macdonald Professor of Physics, McGill University. Louis Carrier, Mercury Press, 1926.

CHAPTER V

THE SPECIAL THEORY OF RELATIVITY

THE theory of relativity has resulted from a combination of the three elements which were called for in a reconstruction of physics: first, delicate experiment; secondly, logical analysis; and thirdly, epistemological considerations. These last played a greater part in the early stages of the theory than in its finished form, and perhaps this is fortunate, since their scope and validity may be open to question, or at least would be but for the successes to which they have led. One may say, broadly, that relativity, like earlier physics, has assumed that when different observers are doing what is called "observing the same phenomenon," those respects in which their observations differ do not belong to the phenomenon, but only those respects in which their observations agree. This is a principle which common sense teaches at an early age. A young child, seeing a ship sailing away, thinks that the ship is continually growing smaller; but before long he comes to recognize that the diminution in size is only "apparent," and that the ship "really" remains of the same size throughout its voyage. In so far as relativity has been inspired by epistemological considerations, they have been of this common-sense kind, and the apparent paradoxes have resulted from the discovery of unexpected differences between our observations and those of other hypothetical observers. Relativity physics, like all physics, assumes the realistic hypothesis, that there are occurrences which different people can observe. For the present, we may ignore epistemology, and proceed to consider relativity simply as theoretical physics. We may also ignore the experimental evidence, and regard the whole theory as a deductive system, since that is the point of view with which we are concerned in Part I.

The most remarkable feature of the theory of relativity, from a philosopher's standpoint, was already present in the special theory: I mean the merging of space and time into space-time. The special theory has now become only an approximation, which is not exactly true in the neighbourhood of matter. But it remains worth understanding, as a stage towards the general theory. Moreover, it does not demand the abandonment of nearly such a large proportion of our common-sense notions as is discarded by the general theory.

Technically, the whole of the special theory is contained in the Lorentz transformation. This transformation has the advantage that it makes the velocity of light the same with respect to any two bodies which are moving uniformly relatively to each other, and, more generally, that it makes the laws of electromagnetic phenomena (Maxwell's equations) the same with respect to any two such bodies. It was for the sake of this advantage that it was originally introduced; but it was afterwards found to have wider bearings and a more general justification. In fact, it may be said that, given sufficient logical acumen, it could have been discovered at any time after it was known that light is not propagated instantaneously. It has grown by this time very familiar—so familiar that I have even seen it quoted (quite correctly) in an advertisement of Fortnum and Mason's. Nevertheless, it is, I suppose, desirable to set it forth. In its simplest form it is as follows:

Suppose two bodies, one of which (S') is moving relatively to the other (S) with velocity v parallel to the x-axis. Suppose that an observer on S observes an event which he judges to have taken place at time t, by his clocks, and in the place whose co-ordinates, for him, are x, y, z. (Each observer takes himself as origin.) Suppose that an observer on S' judges that the event occurs at time t' and that its co-ordinates are x', y', z'. We suppose that at the time when $t=0$ the two observers are at the same place, and also $t'=0$. It would formerly have seemed axiomatic that we should have $t=t'$. Both observers are supposed to employ faultless chronometers, and, of course, to allow for the velocity of light in estimating the time when the event occurs. It would be thought, there fore, that they would arrive at the same estimate as to the time of the occurrence. It would also have been thought that we should have:

$$x'=x-vt.$$

Neither of these, however, is correct. To obtain the correct transformation, put:

$$\beta=\frac{c}{\sqrt{c^2-v^2}},$$

where c is, as always, the velocity of light. Then:

$$\left.\begin{aligned} x'&=\beta(x-vt) \\ t'&=\beta\left(t-\frac{vx}{c^2}\right) \end{aligned}\right\} \quad \ldots\ldots\ldots\ldots\ldots (1).$$

For the other co-ordinates y', z', we still have, as before:

$$y'=y, \quad z'=z.$$

It is the formulæ for x' and t' that are peculiar. These formulæ contain, implicitly, the whole of the special theory of relativity.

The formula for x' embodies the FitzGerald contraction. Lengths on either body, as estimated by an observer on the other, will be shorter than as estimated by an observer on the body on which the lengths are: the longer length will have to the shorter the ratio β. More interesting, however, is the effect as regards time. Suppose that an observer on the body S judges two events at x_1 and x_2 to be simultaneous, and both at time t. Then an observer on S' will judge that they occur at times t_1', t_2', where:

$$t_1'=\beta\left(t-\frac{vx_1}{c^2}\right)$$

$$t_2'=\beta\left(t-\frac{vx_2}{c^2}\right),$$

and therefore:

$$t_1' - t_2' = \beta \frac{v(x_2 - x_1)}{c^2}.$$

This is not zero unless $x_1 = x_2$; thus in general events which are simultaneous for one observer are not simultaneous for the other. We cannot therefore regard space and time as independent, as has always been done in the past. Even the order of events in time is not definite: in one system of co-ordinates an event A may precede an event B, while in another B may precede A. This, however, is only possible if the events are so separated that, no matter how we choose our co-ordinates, light starting from either could not reach the place of the other until after the other had occurred.

The Lorentz transformation yields the result that:

$$c^2t^2 - x^2 = c^2t'^2 - x'^2.$$

Since $y = y'$ and $z = z'$, we have:

$$c^2t^2 - (x^2 + y^2 + z^2) = c^2t'^2 - (x'^2 + y'^2 + z'^2);$$

or, putting r, r' for the distances of the event from the two observers:

$$c^2t^2 - r^2 = c^2t'^2 - r'^2 \quad \ldots\ldots\ldots\ldots\ldots (2).$$

This result is general—*i.e.* given any two reference-bodies in uniform relative motion, if r is the distance between two events according to one system, r' the distance according to the other, and if t, t' are the corresponding time-intervals between the events, equation (2) will always hold. Thus $c^2t^2 - r^2$ represents a physical quantity, independent of the choice of co-ordinates; it is called the square of the "interval" between the two events. There are two cases, according as it is positive or negative. When it is positive, the interval between the events is called "time-like"; when negative, "space-like." In the intermediate case in which it is zero, the events are such that one light-ray can be present at each. In this case, one event might be the seeing of the other. The time-order of two events will be different in different reference-systems when their interval is space-like, but when it is time-like the time-order is the same in all systems, though the magnitude of the time-interval varies.

When the interval between two events is time-like, it is possible for a body to move in such a way as to be present at both events. In that case, the interval is what clocks on that body will show as the time. When the interval between two events is space-like, it is possible for a body to move in such a way that, by its clocks, the two events will be simultaneous; in that case, the interval is what, in relation to that body, appears as their distance. (In these remarks, we are taking the velocity of light as the unit of velocity, which is convenient in relativity theory.) Both these are consequences of the Lorentz transformation. From the first of them it follows that, if two events both happen to me, the time between them as measured by my watch (assuming it to be a good watch) is the " interval " between them, and has still a physical significance. Thus the time that is concerned in psychology is unaffected by relativity, assuming that everything that psychology is concerned with happens, from a physical point of view, in the body of the person whose mental events are being considered. This is an assumption for which grounds will be given at a later stage.

It follows from the ambiguity of simultaneity between distant events that we cannot speak unambiguously of " *the* distance between two bodies at a given time." If the two bodies are in relative motion, a " given time " will be different for the two bodies and different again for other reference-bodies. It follows that such a conception cannot enter into the correct statement of a physical law. On this ground alone, we can conclude that the Newtonian form of the law of gravitation cannot be quite right. Fortunately, Einstein has supplied the necessary correction.

It will be observed that, as a consequence of the Lorentz transformation, the mass of a body will not be the same when it is in motion relatively to the reference-body as when it is at rest relatively to it. The mass of a body is inversely proportional to the acceleration produced in it by a given force, and two reference-bodies in uniform relative motion will give different results for the acceleration of a third body. This is obvious as a consequence of the FitzGerald contraction. The increase of mass with rapid motion was known experimentally before the special theory of relativity had explained it; it is very marked for velocities such as those attained by β-particles (electrons) emitted by radio-active bodies, since these velocities may be as great as 99 per cent. of the velocity of light. This change of mass, like the FitzGerald contraction, seemed strange and anomalous until the special theory of relativity explained it.

One more point is important as showing how easily what seems axiomatic may be false: it concerns the composition of velocities. Suppose three bodies moving uniformly in the same direction: the velocity of the second relatively to the first is v, that of the third relatively to the second is w. What is the velocity of the third relatively to the first? One would have thought it must be $v+w$, but in fact it is:

$$\frac{v+w}{1+\frac{vw}{c^2}}.$$

It will be seen that this $\leqslant c$; if $v=c$ or $w=c$, it is c, otherwise it is less than c. This is an illustration of the way in which the velocity of light plays the part of infinity in relation to material motions.

The special theory set itself the task of making the laws of physics the same relatively to any two co-ordinate systems in uniform rectilinear relative motion. There were two sets of equations to be considered: those of Newtonian dynamics, and Maxwell's equations. The latter are unaltered by a Lorentz transformation, but the former require certain adaptations. These, however, are such as experimental results had already suggested. Thus the solution of the problem in hand was complete, but of course it was obvious from the first that the real problem was more general. There could be no reason for confining ourselves to two co-ordinate systems in uniform rectilinear motion; the problem ought to be solved for any two co-ordinate systems, no matter what the nature of their relative motion. This is the problem which has been solved by the general theory of relativity.

CHAPTER VI

THE GENERAL THEORY OF RELATIVITY

THE general theory of relativity has a much wider sweep than the special theory, and a greater philosophical interest, apart from the one matter of the substitution of space-time for space and time. The general theory demands an abandonment of all direct relations between distant events, the relations upon which space-time depends being primarily confined to very small regions, and only extended, where they can be extended, by means of integration. All the old apparatus of geometry —straight lines, circles, ellipses, etc.—is gone. What belongs to *analysis situs* remains, with certain modifications; and there is a new geometry of geodesics, which has come from Gauss's study of surfaces by way of Riemann's inaugural dissertation. Geometry and physics are no longer distinct, so long as we are not considering the parts of physics which introduce atomicity, such as electrons, protons, and quanta. Perhaps even this exception may not long remain. There are parts of physics which, so far, lie outside the general theory of relativity, but there are no parts of physics to which it is not in some degree relevant. And its importance to philosophy is perhaps even greater than its importance to physics. It has, of course, been seized upon by philosophers of different schools as affording support to their respective nostrums; St. Thomas, Kant, and Hegel are claimed to have anticipated it. But I do not think that any of the philosophers who make these suggestions have taken the trouble to understand the theory. For my part, I do not profess to know exactly what its philosophical consequences will prove to be, but I am convinced that they are far-reaching, and quite different from what they seem to philosophers who are ignorant of mathematics.

In the present chapter, I wish to consider Einstein's theory without any regard to its philosophical implications, simply as a logical system. The system starts by assuming a four-dimensional manifold having a definite order. The form which this assumption takes is somewhat technical: it is assumed that, when we have what might be called an ordinary set of co-ordinates—*e.g.* those which would naturally be employed in Newtonian astronomy—there are certain transformations of these co-ordinates which are legitimate, and certain others which are not. Those which are legitimate are those which transform infinitesimal distances into infinitesimal distances. This means to say that the transformations must be continuous. Perhaps what is assumed may be stated as follows: Given a set of points $p_1, p_2, p_3, \ldots$ whose co-ordinates tend towards a limiting set which is the co-ordinates of a point p, then in any new legitimate co-ordinate system those points $p_1, p_2, p_3, \ldots$ must have co-ordinates tending to a limiting set which is the co-ordinates of p in the new system. This means that certain relations of order among the co-ordinates represent properties of the points of space-time, and are presupposed in the assignment of co-ordinates. The accurate statement of what is involved can only be made in terms of limits, but the correct meaning is conveyed by saying that neighbouring points must have neighbouring co-ordinates. The exact nature of the ordinal presuppositions of a relativistic co-ordinate system will occupy us in a later chapter; for the present I merely wish to emphasize that the space-time manifold, in the general theory of relativity, has an order which is not arbitrary, and which is reproduced in any legitimate co-ordinate system. This order, it is important to realize, is *purely* ordinal, and does not involve any metrical element. Nor is it derivable from the metrical relations of points which are afterwards introduced in the theory—*i.e.* from " intervals."

The points of space-time have, of course, no duration as well as no spatial extension. It is generally assumed that several events may occupy the same point; this is involved in the conception of the intersection of world-lines. I think it may also be assumed that one event may extend over a finite extent of space-time, but on this point the theory is silent, so far as I know. I shall myself, in a later chapter, deal with the construction of points as systems of events, each of which events has a finite extension; this is a subject which has been especially treated by Dr Whitehead, but I shall suggest a method somewhat different from his. So long as we confine ourselves to the theory of relativity, it is not necessary to consider whether events have a finite extension, though I think it is necessary to assume that two events may both occupy the same point of space-time. Even on this, however, there is a certain vagueness in the authoritative expositions, which is due mainly to the large scale of the phenomena with which the theory is principally concerned. Sometimes it would seem as if the whole earth counted as a point; certainly one physical laboratory does so in the practice of writers on relativity. On occasion, Professor Eddington considers an area of 9×10^{10} square kilometres to be an infinitesimal of the second order. The fact that such a view is appropriate in discussions of relativity makes it unnecessary to be precise as to what is meant by saying that two events occupy the same point, or that two world-lines intersect. For the present I shall assume that this is possible in a strict sense; my reasons will be given in a later chapter.

It is assumed that every point of space-time can have four real numbers assigned to it, and conversely that any four real numbers (at any rate within certain limits) are the co-ordinates of a point. This amounts to the assumption that the number of points is $2^{\aleph_0}$, where $\aleph_0$ is the number of finite integers; that is to say, the number of points is the number of the Cantorian continuum. Every class of $2^{\aleph_0}$ terms is the field of various multiple relations which arrange the class in a four-dimensional continuum—or an *n*-dimensional continuum, for that matter. But we require a little more than this. Of all the ways of arranging the points of space-time in a four-dimensional continuum, there is only one that has physical significance; the others exist only for mathematical logic. That means that there must be among points relations derivable from an empirical basis, which generate a four-dimensional continuum. These will be the ordinal relations spoken of in the last paragraph but one. We assume, therefore, that these ordinal relations generate a continuum, and that co-ordinates are so assigned that neighbouring points have neighbouring co-ordinates. More exactly the co-ordinates of the limit of a set of points are the limits of the co-ordinates of the set. This is not a law of nature, but a prescription as to the manner in which co-ordinates are assigned. It leaves great latitude, but not complete latitude. It allows any system of co-ordinates to be replaced by another system in which the new co-ordinates are any continuous functions of the old co-ordinates, but it excludes discontinuous functions.

We now assume that any two neighbouring points have a metrical relation, called their "interval," whose square is a quadratic function of the differences of their co-ordinates. This is a generalization of the theorem of Pythagoras, which has come by way of Gauss and Riemann. It will be worth while to consider the historical development for a moment.

By the theorem of Pythagoras, if two points in a plane have co-ordinates (x_1, y_1), (x_2, y_2), and s is their distance apart:

$$s^2 = (x_2 - x_1)^2 + (y_2 - y_1)^2.$$

By an immediately obvious extension, if two points in space

have co-ordinates (x_1, y_1, z_1), (x_2, y_2, z_2), their distance apart is s, where:

$$s^2 = (x_2 - x_1)^2 + (y_2 - y_1)^2 + (z_2 - z_1)^2.$$

If the distance apart is small, we write dx, dy, dz for $x_2 - x_1$, $y_2 - y_1$, $z_2 - z_1$, and ds for s; thus:

$$ds^2 = dx^2 + dy^2 + dz^2.$$

Gauss considered a problem concerned with surfaces, which arises naturally out of the above. On a surface, the position of a point can be fixed by two co-ordinates, which need not involve reference to anything outside the surface. Thus on the earth position is fixed by latitude and longitude. Suppose u and v are two such co-ordinates which fix position on a surface. Then in general we shall not have:

$$ds^2 = du^2 + dv^2$$

for the distance between neighbouring points; in general, we cannot get a formula of this kind however we may define u and v. We can get a formula of this kind on a cylinder or a cone, and generally on what are called "developable" surfaces, but not, *e.g.*, on a sphere. The general formula takes the shape:

$$ds^2 = Edu^2 + 2Fdudv + Gdv^2,$$

where E, F, G are in general functions of u and v, not constants. Gauss showed that there are certain functions of E, F, G which have the same value however the co-ordinates u and v may be defined; these functions express properties of the surface, which can theoretically be discovered by measurements carried out on the surface, without reference to external space.

Riemann extended this method to space. He supposed that the theorem of Pythagoras may be not exact, and that the correct formula for the distance between two points may be such as results from Gauss's formula by adding another variable. He showed that this supposition could be made the basis of non-Euclidean geometry. The whole subject of non-Euclidean geometry remained, however, without visible relevance to physics until it was utilized in Einstein's theory of gravitation, which results from the combination of Riemann's ideas with the substitution of space-time "interval" for distance in space and time, which had already been made in the special theory of relativity.

In the special theory of relativity, as we saw, the interval between two space-time points, one of which is the origin, is s, where:

$$s^2 = x^2 + y^2 + z^2 - c^2t^2,$$

if the interval is space-like, and:

$$s^2 = c^2t^2 - x^2 - y^2 - z^2,$$

if the interval is time-like. In practice, the latter form is always taken. Any system of co-ordinates allowed by the special theory gives the same value for the interval between two given space-time points. But we are now allowing much greater latitude in the choice of co-ordinates, and we are assuming that the special theory represents only an approximation, being not strictly true except in the absence of a gravitational field. We still assume that, for small distances, there is a quadratic function of the co-ordinate differences which has a physical significance, and has the same value however the co-ordinates may be assigned, subject to the condition of continuity already explained. That is, if x_1, x_2, x_3, x_4 are the co-ordinates of a point, and $x_1 + dx_1$, $x_2 + dx_2$, $x_3 + dx_3$, $x_4 + dx_4$ are the co-ordinates of a neighbouring point, we assume that there is a quadratic function:

$$\Sigma g_{\mu\nu} dx_\mu dx_\nu \quad (\mu, \nu = 1, 2, 3, 4),$$

which has the same value however the co-ordinates may be assigned; we then define ds as the "interval" between the two neighbouring points. The $g_{\mu\nu}$'s will be functions of the co-ordinates (in general not constants), and for convenience we take $g_{\mu\nu} = g_{\nu\mu}$. Just as Gauss was able to deduce the geometry of a surface from his formula, so we can deduce the geometry of space-time from our formula. But as we include time, our geometry is not merely geometry, but physics; in other words, it combines history with geography.

At a great distance from matter, the special theory will still be true, and therefore space will be Euclidean, since, if we put $dt=0$, the special theory gives the Euclidean formula for distance. The neighbourhood of gravitating matter is shown by a non-Euclidean character of the region concerned. This, however, requires some preliminary explanations, more especially an explanation of the method of tensors, which will form the subject of the next chapter.

Everything in the general theory of relativity is dependent upon the existence of the above formula for ds^2. The formula itself is of the nature of an empirical generalization; no *a priori* justification for it is suggested. It is a generalization of the theorem of Pythagoras, which could formerly be proved. But the proof rested upon Euclid's axioms, which there is no reason to regard as exactly true. More than that, there is difficulty in assigning a meaning to his fundamental concepts, such as the "straight" line. The old geometry assumed a static space, which it could do because space and time were supposed to be separable. It is natural to think of motion as following a path in space which is there before and after the motion: a tram moves along pre-existing tram-lines. This view of motion, however, is no longer tenable. A moving point is a series of positions in space-time; a later moving point cannot pursue the "same" course, since its time co-ordinate is different, which means that, in another equally legitimate system of co-ordinates, its space co-ordinates also will be different. We think of a tram as performing the same journey every day, because we think of the earth as fixed; but from the sun's point of view, the tram never repeats a former journey. "We cannot step twice into the same rivers," as

Heraclitus says. It is thus obvious that, in place of Euclid's static straight line, we shall have to substitute a movement having some special property defined in terms of space-time, not of space. The movement required is a " geodesic," concerning which we shall have more to say later.

In relativity theory, distant space-time points have only such relations as can be obtained by integration from the relations of neighbouring points. Since the distance between two points is always finite, what we call a relation between neighbouring points is not really a relation between points at all, but is a limit, like a velocity. Only the language of the calculus can express accurately what is meant. One might say, speaking pictorially, that the notion of " interval " is concerned with what, at each point, is *tending* to happen, although we cannot say that this will actually happen, because before any assigned point is reached something may have occurred to cause a diversion. This is, of course, the case with velocity. From the fact that, at a given instant, a body is moving in a given direction with a given velocity, we can infer nothing whatever as to where the body will be at another assigned instant, however near to the first. To infer the path of a body from its velocity, we must know its velocity throughout a finite time. Similarly the formula for interval characterizes each separate point of space-time. To obtain the interval between one point and another, however near together, we must specify a route, and integrate along that route. As we shall see, however, there are routes which may be called " natural "—namely, geodesics. It is only by means of them that the notion of interval can be profitably extended to the relations of points at a finite distance from each other.

CHAPTER VII

THE METHOD OF TENSORS

THE method of tensors contains the answer to a question which is rendered urgent by the arbitrary character of our co-ordinates. How can we know whether a formula expressed in terms of our co-ordinates expresses something which describes the physical occurrences, and not merely the particular co-ordinate system which we happen to be employing? A striking example of the mistakes that are possible in this respect is afforded by simultaneity. Suppose we have two events, whose co-ordinates, in the system we are employing, are (x, y, z, t) and (x', y', z', t)—*i.e.* their time co-ordinates are the same. Before the special theory of relativity everybody would have asserted that this represented a physical fact about the two events—namely, that they are simultaneous. Now we know that the fact concerned is one which also involves mention of the co-ordinate system—that is to say, it is not a relation between the two events only, but between them and the body of reference. But this is to speak the language of the special theory. In the general theory, our co-ordinates may have no important physical significance, and a pair of events which have one co-ordinate identical need not have any intrinsic physical property not possessed by other pairs of events. In practice, there must be *some* principle on which co-ordinates are assigned, and this principle must have *some* physical significance. But we might, for instance, measure time by the worst clock ever made, provided it only went wrong and did not actually stop. And we might use a certain worm as our unit of length, disregarding the "FitzGerald contraction" to which motion subjects him. In that case, if we say that there was unit distance between two events which both occurred at a certain instant, we shall be making a complicated comparison between the events, a bad clock, and a certain worm—that is to say, we shall be making a statement which depends upon our co-ordinate system. We want to discover a sufficient, if not necessary, condition which, if fulfilled, insures that a statement in terms of co-ordinates has a meaning independent of co-ordinates. The difference is more or less analogous to that, in ordinary language, between linguistic statements and statements which (as is usually the case) are about what words mean. If I say "strength is a desirable quality," my statement can be put into French or German without change of meaning. But if I say "strength is a word containing seven consonants and only one vowel," my statement becomes false if translated into French or German. Now in physics co-ordinates are analogous to words, with the difference that it is much harder to distinguish "linguistic" statements from others. This is what the method of tensors undertakes to do.

It does not seem possible to state the method of tensors in untechnical language; I am afraid that those philosophers who have not thought it worth while to learn the calculus cannot hope to understand it. Perhaps in time some simple way of explaining it may be found, but none has been found so far.*

Suppose we have a vector quantity whose components are A^1, A^2, A^3, A^4. (Here 1, 2, 3, 4 play the part of suffixes, not of exponents denoting powers.) It happens in certain cases that, if we transform to any other co-ordinates x'_1, x'_2, x'_3, x'_4, which are continuous functions of the old co-ordinates x_1, x_2, x_3, x_4, we shall have, as the components of the vector in the new co-ordinates, A'^1, A'^2, A'^3, A'^4, where:

$$A'^1 = \frac{\partial x_1'}{\partial x_1}A^1 + \frac{\partial x_1'}{\partial x_2}A^2 + \frac{\partial x_1'}{\partial x_3}A^3 + \frac{\partial x_1'}{\partial x_3}A^1$$

* For what follows see Eddington, *Mathematical Theory of Relativity*, chap. ii., Cambridge, 1924.

with similar formulæ for A'^2, A'^3, A'^4. When this happens, the vector in question is called *contravariant*. The simplest example is (dx_1, dx_2, dx_3, dx_4). Except in this one case, the " contravariant " property is symbolized by the upper position of the suffix.

Again we may have a vector, whose components are A_1, A_2, A_3, A_4, which is transformed according to the law :

$$A'_1 = \frac{\partial x_1}{\partial x_1'}A_1 + \frac{\partial x_2}{\partial x_1'}A_2 + \frac{\partial x_3}{\partial x_1'}A_3 + \frac{\partial x_4}{\partial x_1'}A_4$$

with similar formulæ for A'_2, A'_3, A'_4. Such a vector is called *covariant*. The simplest example is the vector whose components are:

$$\frac{\partial \varphi}{\partial x_1}, \frac{\partial \varphi}{\partial x_2}, \frac{\partial \varphi}{\partial x_3}, \frac{\partial \varphi}{\partial x_4},$$

where φ is some function which has a fixed value at each point, independently of the co-ordinate system.

It is obvious that, if we have two contravariant vectors A and B whose components are equal in one system of co-ordinates, then their components are equal in any system of co-ordinates; and the same applies to two covariant vectors A and B. This follows at once from the above rules of transformation. Thus an equality of two contravariant vectors, or of two covariant vectors, when it occurs, is a fact independent of the co-ordinate system. It is, in fact, a tensor equation of the simplest kind.

The general definition of a " tensor " is a generalization of those of contravariant and covariant vectors. Instead of a vector with only four components, we may have a quantity with sixteen components:

$$A_{11}, A_{12}, A_{13}, A_{14}, A_{21}, A_{22}, A_{23}, A_{24},$$
$$A_{31}, A_{32}, A_{33}, A_{34}, A_{41}, A_{42}, A_{43}, A_{44}.$$

Such a quantity may be denoted by " $A_{\mu\nu}$," where it is understood that μ and ν can each take all values from 1 to 4.

Similarly we may have a quantity with sixty-four components, A_{111}, A_{112}, etc.; such a quantity may be denoted by "$A_{\mu\nu\sigma}$," where μ and ν and σ can each take all values from 1 to 4. Such quantities are called "tensors" if they obey laws of transformation analogous to those of contravariant and covariant vectors. Thus a contravariant tensor with sixteen components, which is written "$A^{\mu\nu}$," is one which satisfies the rule:

$$A'^{11}=\left(\frac{\partial x'_1}{\partial x_1}\right)^2 A^{11}+\left(\frac{\partial x'_1}{\partial x_2}\right)^2 A^{22}+\ldots+\frac{\partial x'_1}{\partial x_1}\frac{\partial x'_1}{\partial x_2}A^{12}+\ldots$$

with similar equations for the other components—*e.g.*:

$$A'^{12}=\frac{\partial x'_1}{\partial x_1}\frac{\partial x'_2}{\partial x_1}A^{11}+\ldots+\frac{\partial x'_1}{\partial x_1}\frac{\partial x'_2}{\partial x_2}A^{12}+\ldots$$

These equations are comprised in:

$$A'^{\mu\nu}=\sum_{\alpha,\beta}\frac{\partial x'_\mu}{\partial x_\alpha}\frac{\partial x'_\nu}{\partial x_\beta},$$

where α, β are to take all values from 1 to 4. Similarly a covariant tensor with sixteen components, written "$A^{\mu\nu}$," is one which is transformed according to the rule:

$$A'_{\mu\nu}=\sum_{\alpha,\beta}\frac{\partial x_\alpha}{\partial x'^\mu}\frac{\partial x_\beta}{\partial x'_\nu}A_{\alpha\beta},$$

and a mixed tensor, written A^ν_μ, is one which satisfies the rule:

$$A'^\nu_\mu=\sum_{\alpha,\beta}\frac{\partial x_\alpha}{\partial x'_\mu}\frac{\partial x'_\nu}{\partial x_\beta}A^\beta_\alpha.$$

There is no difficulty in extending these definitions to any number of suffixes. It is obvious, as in the case of contravariant and covariant vectors, that if two tensors of the same kind are equal in one system of co-ordinates they are equal in any system of co-ordinates, so that tensor equations express conditions which are independent of the choice of co-ordinates. For this reason it is necessary to express all the general laws of physics as tensor equations; if this cannot be done, the law concerned must be wrong, and must require such correction as will enable it to be expressed as a tensor equation. The law of gravitation is the most noteworthy example of this; but perhaps the conservation of energy is scarcely less noteworthy.*

It seems natural to suppose that it would be possible to develop a less indirect method of expressing physical laws than that afforded by the method of tensors, which is perhaps a consequence of the historical development of physics. Originally, in physics, the co-ordinates were intended to express physical relations between the event concerned and the origin. Three of the co-ordinates were lengths, which, it was thought, could be ascertained by measurement with a rigid rod. The fourth was a time, which could be measured by a chronometer. There were difficulties, however, which the progress of physics made increasingly evident. So long as the earth could be regarded as motionless, axes fixed relatively to the earth and clocks which remained on the surface of the earth seemed to suffice. It was possible to disregard the facts that no body is quite rigid and no clock quite accurate, because the system of physical laws suggested by the choice of the most rigid bodies and the most accurate clocks could be used to estimate the departure of these instruments from strict constancy, and the results were on the whole self-consistent. But in astronomical problems, including that of the tides, the earth could not be treated as fixed. It was necessary to Newtonian dynamics that the axes should not have any acceleration, but it resulted from the law of gravitation that any material axes must have some acceleration. The axes, therefore, became ideal structures in absolute space; actual measurements with actual rods could only approximate to the results which would have followed if we could have used unaccelerated axes. This difficulty was not the most serious: the worst trouble was concerned with absolute

* See Eddington, *op. cit.*, p. 134.

acceleration. Then came the experimental discovery of the facts which led to the special theory of relativity: the variation of length and mass with velocity, and the constancy of the velocity of light *in vacuo* no matter what body was used to define the co-ordinates. This set of difficulties was solved by the special theory of relativity, which showed that equivalent results come from employing as reference-body any one of a set of bodies in uniform rectilinear motion. This, however, only achieved what Galileo and Newton thought they had achieved. It included electromagnetic phenomena within the scope of relativity as regards velocities, but it was clearly necessary to extend relativity to accelerations, and when this was done, co-ordinates ceased to have the clear physical meaning they had formerly possessed. It is true that, even in the general theory, a co-ordinate, in any system which can actually be used, will always have *some* physical significance, but its significance is trivial and complicated, not, as before, important and simple.

It is natural to ask: Could we not dispense with co-ordinates altogether, since they have become little more than conventional names systematically assigned? Perhaps this will become possible in time, but at present the necessary mathematics is lacking. We wish, for example, to be able to differentiate, and we cannot differentiate a function unless its arguments and values are numbers. This is not due to what might seem the more difficult parts of the definition of a differential. We can define for a non-numerical function the limit (if it exists) of a function for a given argument, and also the four limits which exist more frequently—viz. the maximum and minimum for approaches from above and below; we can also define a "continuous" non-numerical function. (See *Principia Mathematica*, *230-*234.) What, so far, has not been defined, except for numbers, is a fraction. Now $\frac{dy}{dx}$ is the limit of a fraction; thus, although we can generalize the notion of a limit, we cannot at present generalize $\frac{dy}{dx}$, because we cannot generalize the notion of a fraction. It seems clear *a priori* that, since differentiation of co-ordinates is physically useful even when the quantitative value of the co-ordinates is conventional, there must be some process, of which differentiation is a special numerical form, which can be applied wherever we have continuous functions, even when they are non-numerical. To define such a process is a problem in mathematical logic, probably soluble, but hitherto unsolved. If it were solved, it might become possible to avoid the elaborate and round-about process of assigning co-ordinates and then treating almost all their properties as irrelevant, which is what is done when the method of tensors is employed.

There are, it is true, certain numbers which are important in the new geometry: they are those giving the measure of intervals. But, as we have already seen, two points at a finite distance apart do not have an unambiguous interval; and any two points are at a finite distance apart. The numbers involved in the notion of interval are not finite distances, but numbers derivable from the sixteen coefficients $g_{\mu\nu}$ involved in the formula for ds^2 in the previous chapter. These coefficients themselves depend upon the co-ordinate system, but ds^2 does not. We cannot develop this theme until we have considered geodesics; it is from them that we must derive the numbers which have, in the new geometry, the same sort of physical importance as co-ordinates were supposed to have in the old. These numbers will be the integrals of ds taken along certain geodesics. But, unlike lengths in the old metrical geometry, they are geometrically insufficient. To avoid irrelevant complications, we may illustrate this insufficiency by considering the special theory.

The most obvious example of the failure of interval to

constitute a geometry is derived from consideration of light-rays. The interval between two events which are parts of the same light-ray is zero. Suppose now that a light-ray starts from an event A, and arrives at an event B; at the moment when it reaches B, another light-ray starts from B and reaches C. Then the interval between A and B is zero, that between B and C is zero, but that between A and C may have any time-like magnitude. Euclid proved that two sides of a triangle are together greater than the third side, and was criticized on the ground that this proposition was evident even to asses. But in relativity geometry this proposition is false. In our triangle ABC, AB and BC are zero, while AC may have any finite magnitude.

Again, the events which are parts of a single light-ray have a definite time-order, in spite of the fact that the interval between any two of them is zero. This appears as follows. Suppose a light-ray proceeds from the sun to the moon and is thence reflected to the earth: it reaches the earth later than a direct ray which left the sun at the same time. There is therefore a definite sense in saying that the ray reached the moon later than it left the sun—*i.e.* we can say that the ray went from the sun to the moon, not from the moon to the sun. Generalizing, we may say: If A and B are part of one light-ray, and light-rays from A and B, distinct from the previous light-ray, contain events C, C' whose interval is time-like, then the time-order of C, C' is the same whatever these new light-rays may be—*i.e.* we shall have always C before C', or always C' before C. In the first case, we say that the "sense" of the ray is from A to B; in the second, from B to A. This illustrates the difficulties which would arise if we were to attempt to found our geometry on interval alone. We must also take account of the purely ordinal properties of the space-time manifold. These properties give a wide separation between the departure of a light-ray from the sun and its arrival on the earth, although the " interval " between these two events is zero.

Reverting now to the method of tensors and its possible eventual simplification, it seems probable that we have an example of a general tendency to over-emphasize numbers, which has existed in mathematics ever since the time of Pythagoras, though it was temporarily less prominent in later Greek geometry as exemplified in Euclid. Euclid's theory of proportion does not, of course, dispense with numbers, since it uses " equimultiples "; but at any rate it requires only integers, not irrationals. Owing to the fact that arithmetic is easy, Greek methods in geometry have been in the background since Descartes, and co-ordinates have come to seem indispensable. But mathematical logic has shown that number is logically irrelevant in many problems where it formerly seemed essential, notably mathematical induction, limits, and continuity. A new technique, which seems difficult because it is unfamiliar, is required when numbers are not used; but there is a compensating gain in logical purity. It should be possible to apply a similar process of purification to physics. The method of tensors first assigns co-ordinates, and then shows how to obtain results which, though expressed in terms of co-ordinates, do not really depend upon them. There must be a less indirect technique possible, in which we use no more apparatus than is logically necessary, and have a language which will only express such facts as are now expressed in the language of tensors, not such as depend upon the choice of co-ordinates. I do not say that such a method, if discovered, would be preferable in practice, but I do say that it would give a better expression of the essential relations, and greatly facilitate the task of the philosopher. In the meantime, the method of tensors is technically delightful, and suffices for mathematical needs.

CHAPTER VIII

GEODESICS

THE importance of geodesics arises through the law that, in the general theory of relativity, a particle not subject to constraints moves in a geodesic. But let us first consider what a geodesic is.

An adventurous pedestrian in the Alps may wish to go from a place in one valley to a place in another by the shortest route—*i.e.* the shortest compatible with remaining all the time on the earth's surface. He cannot determine the shortest route by looking at a large-scale map and drawing a straight line between the two places, for if this line involves a greater average gradient than another it may be longer, in distance as well as in time, than another route which slopes gradually to the head of a pass and then down again. What the traveller is seeking is a " geodesic "—*i.e.* the shortest line that can be drawn on the earth's surface between the two points. In the absence of hills—*e.g.* on the sea—the shortest route is by a great circle. On complicated surfaces, geodesics may become very complicated curves. The definition is not exactly " the shortest route between two points." The definition is that the distance along a geodesic from any one of its points to any other must be " stationary "—*i.e.* such that either all very slightly different paths are longer, or all very slightly different paths are shorter. This means that, for small variations of path, the first-order change of length is zero. In effect, in the ordinary geometry of surfaces the geodesic distance is a minimum, and in relativity theory it is a maximum. This is not so great a difference as it may seem to the non-mathematical reader, since the geodesic distance concerned in relativity theory is more analogous to what would ordinarily count as lapse of time than to what would ordinarily count as distance in space.

Let us try to make the matter a little more concrete. The earth, in its annual revolution, travels from place to place in space-time; between the positions of Greenwich Observatory on two occasions six months apart, there is a certain interval. From the point of view of an observer in the sun, the interval would formerly have been divided into two parts—namely, six months and about 186,000,000 miles. But from the point of view of the observer at Greenwich there is only one interval—namely, time—since the place concerned is the same on both occasions. Given a clock which travels without constraint from one point of space-time to another, the interval between these two points is what that clock registers as the time between them. I say that if a clock were constrained to travel by some other slightly different route, so as to be present at Greenwich Observatory on two occasions six months apart, but absent from the earth in the meantime, the time which that clock would register as having been taken by its journey would be less than six months. The interval between distant points is not, like distance in geometry, something which can be defined independently of the route chosen. The interval must be obtained by integration along a specified route, and a geodesic route is one which makes the interval greater than it is by any slightly different route. The time between two given events at which a man is present seems less if he has spent the intervening time in rapid travel than if he has let himself drift passively; this is a sort of law of cosmic boredom. All bodies, left to themselves, choose the course which is at each moment the most boring, in the sense that it makes the time between two given events seem longest. However, it is time to have done with these irrelevancies, and return to seriousness.

Since the small interval ds is independent of the co-ordinates, a geodesic also is independent of them. We can easily obtain the differential equations which a geodesic must satisfy, and these equations must be satisfied by the same lines whatever system of co-ordinates we are employing. From a given point, goedesics start in all directions. Some of these are the paths of freely moving particles; others are not. The law that the path of a particle is a geodesic does not tell us quite as much as it seems to do, since it is only by observation of the motions of bodies that we discover what paths are geodesics. Assuming that the orbit of the earth is a geodesic, we can draw inferences as to the nature of the formula for ds^2 in the sun's gravitational field. For we have no *a priori* knowledge about the coefficients $g_{\mu\nu}$ which appear in the formula for ds^2; their values are to be deduced from observation. What we can say is that it is possible, compatibly with observed facts, so to determine the $g_{\mu\nu}$ that the path of a body in a gravitational field shall be a geodesic. In fact, we get in this way a more accurate representation of the facts than we got from the Newtonian law, but the observable differences between the two are few and minute.

Although the new law of gravitation and the old do not lead to very different results—as, indeed, they could not, since the old law accorded closely with observed facts—yet the difference in the ideas involved is very great. A planet, in the new theory, is moving freely, whereas in the old theory it was subject to a central force directed towards the sun. In the old theory, the planet moved in an ellipse; in the new theory, it moves in the nearest possible approach to a straight line—to wit, a geodesic. In the old theory, the sun was like a despotic government, emitting decrees from the metropolis; in the new, the solar system is like the society of Kropotkin's dreams, in which everybody does what he prefers at each moment, and the result is perfect order. The odd thing is that, as far as observation goes, the difference between these two theories is exceedingly minute. To the plain man, it would seem impossible to reconcile the statement that the earth moves in an ellipse with the statement that it moves in a sort of straight line, however queer the sort may be. And yet almost the whole of the difference between these two statements is a matter of convention. It is possible to adhere to Euclidean space even now; this requires a different way of stating Einstein's law of gravitation, but does not demand the rejection of anything that has been proved true. Dr Whitehead considers this plan preferable to Einstein's. What may be called the new orthodoxy, per contra, is set forth by Professor Eddington. It will be worth while to consider the point at issue between them.

Professor Eddington says (*op. cit.*, p. 37):

" Suppose that an observer has chosen a definite system of space co-ordinates and of time-reckoning (x_1, x_2, x_3, x_4), and that the geometry of these is given by:

$$ds^2 = g_{11}dx_1^2 + g_{22}dx_2^2 + \ldots + 2g_{12}dx_1dx_2 \quad \ldots\ldots\ldots (16 \cdot 1).$$

Let him be under the mistaken impression that the geometry is:

$$ds_0^2 = -dx_1^2 - dx_2^2 - dx_3^2 + dx_4^2 \quad \ldots\ldots\ldots\ldots (16 \cdot 2)$$

—that being the geometry with which he is most familiar in pure mathematics. We use ds_0 to distinguish his mistaken value of the interval. Since intervals can be compared by experimental methods, he ought soon to discover that his ds_0 cannot be reconciled with observational results, and so realize his mistake. But the mind does not so readily get rid of an obsession. It is more likely that our observer will continue in his opinion, and attribute the discrepancy of the observations to some influence which is present and affects the behaviour of his test-bodies. He will, so to speak, introduce a supernatural agency which he can blame for the consequences of his mistake. Let us examine what name he would apply to this agency.

"Of the four test-bodies considered the moving particle is in general the most sensitive to small changes of geometry, and it would be by this test that the observer would first discover discrepancies. The path laid down for it by our observer is:

$$\int ds_0 \text{ is stationary}$$

—*i.e.* a straight line in the co-ordinates (x_1, x_2, x_3, x_4). The particle, of course, pays no heed to this, and moves in the different track:

$$\int ds \text{ is stationary.}$$

Although apparently undisturbed it deviates from 'uniform motion in a straight line.' The name given to any agency which causes deviation from uniform motion in a straight line is *force* according to the Newtonian definition of force. Hence the agency invoked through our observer's mistake is described as a 'field of force.'

"The field of force is not always introduced by inadvertence, as in the foregoing illustration. It is sometimes introduced deliberately by the mathematician—*e.g.* when he introduces the centrifugal force. There would be little advantage and many disadvantages in banishing the phrase 'field of force' from our vocabulary. We shall therefore regularize the procedure which our observer has adopted. We call (16·2) the *abstract geometry* of the system of co-ordinates (x_1, x_2, x_3, x_4); it may be chosen arbitrarily by the observer. The *natural geometry* is (16·1).

"*A field of force represents the discrepancy between the natural geometry of a co-ordinate system and the abstract geometry arbitrarily ascribed to it.*

"A field of force thus arises from an attitude of mind. If we do not take our co-ordinate system to be something different from that which it really is, there is no field of force."

It is not quite clear why the man who uses forces with a conventional geometry should be regarded as making a "mistake," while the man who says that free particles travel in geodesics, and to justify himself has a queer geometry, is thought to be saying something substantially more accurate.

It is true that we must not conceive " force " as an actual agency, as the older mechanics did; it is merely part of the method of describing how bodies move. But as soon as this is recognized, it is a mere question of convenience whether we speak of forces or not. Let it be conceded that the method of the general theory of relativity is better from a logico-æsthetic point of view; I do not see, however, why we should regard it as any more " true." I am not considering, at the moment, the fact that Einstein's law of gravitation gives a slightly more accurate picture of the phenomena than Newton's, since this is not really relevant to the particular point at issue.

Let us now consider Dr Whitehead's view, which is, on this point, the opposite of Professor Eddington's. In the Preface to *The Principle of Relativity*,* he says:

" As the result of a consideration of the character of our knowledge in general, and of our knowledge of nature in particular, . . . I deduce that our experience requires and exhibits a basis of uniformity, and that in the case of nature this basis exhibits itself as the uniformity of spatio-temporal relations. This conclusion entirely cuts away the casual heterogeneity of these relations which is the essential of Einstein's later theory. It is this uniformity which is essential to my outlook, and not the Euclidean geometry which I adopt as lending itself to the simplest exposition of the facts of nature. I should be very willing to believe that each permanent space is either uniformly elliptic or uniformly hyperbolic, if any observations are more simply explained by such a hypothesis. It is inherent in my theory to maintain the old division between physics and geometry. Physics is the science of the contingent relations of nature, and geometry expresses its uniform relatedness."

Again, in discussing the structure of space-time, he says (*ib.*, p. 29):

* Cambridge, 1922, p. v.

"The structure is uniform because of the necessity for knowledge that there be a system of uniform relatedness, in terms of which the contingent relations of natural factors can be expressed. Otherwise we can know nothing until we know everything."

And on p. 64:

"Though the character of time and space is not in any sense *a priori*, the essential relatedness of any perceived field of events to all other events requires that this relatedness of all events should conform to the ascertained disclosure derived from the limited field. For we can only know that distant events are spatio-temporally connected with the events immediately perceived by knowing what these relations are. In other words, these relations must possess a systematic uniformity in order that we may know of nature as extending beyond isolated cases subjected to the direct examination of individual perception. . . . This doctrine leads to the rejection of Einstein's interpretation of his formulæ, as expressing a casual heterogeneity of spatio-temporal warping, dependent upon contingent adjectives."

Thus whereas Eddington seems to regard it as necessary to adopt Einstein's variable space, Whitehead regards it as necessary to reject it. For my part, I do not see why we should agree with either view: the matter seems to be one of convenience in the interpretation of formulæ. Nevertheless, Dr Whitehead's arguments deserve careful examination.

The main force of the above passages is epistemological: the question involved is the Kantian one, How is knowledge possible? I do not wish to deal with this question in its general form. But without going into theory of knowledge, there is what may be called a common-sense answer. Einstein enables us to predict what in fact can be predicted about astronomical occurrences, and that seems all that ought to be demanded of him. Dr Whitehead objects to the "casual" heterogeneity of space-time in Einstein's system. In a sense, this adjective is justified, since the character of space-time in any region depends upon circumstances which can only be ascertained empirically—namely, the distribution of matter in the neighbourhood. But in another sense the adjective is not justified, since Einstein's law of gravitation gives the rule according to which space-time is affected by the neighbourhood of matter. To say that we cannot, by the help of this rule, know in advance the geometry of a region we have not explored, seems an insufficient objection, since we also cannot know what astronomical occurrences will take place unless we know the distribution of matter. Einstein, like other people, assumes the permanence of matter; this is a point to be considered in another connection, but it has no particular relevance to the present issue. The way the heavenly bodies move depends upon the distribution of matter in their neighbourhood, which is, in Dr Whitehead's phrase, "casual." Even by assuming Euclidean geometry we cannot make astronomical predictions unless we assume that we know the important facts about the distribution of matter in the region concerned. Whether we put the consequences of these facts into our geometry or not does not seem to make any real difference to the possibility of physical knowledge. In all theoretical physics, there is a certain admixture of facts and calculations; so long as the combination is such as to give results which observation confirms, I cannot see that we can have any *a priori* objection. Dr Whitehead's view seems to rest upon the assumption that the principles of scientific inference ought to be in some sense "reasonable." Perhaps we all make this assumption in one form or another. But for my part I should prefer to infer "reasonableness" from success, rather than set up in advance a standard of what can be regarded as credible.

I do not therefore see any ground for rejecting a variable geometry such as Einstein's. But equally I see no ground for supposing that the facts necessitate it. The question is, to my mind, merely one of logical simplicity and comprehensiveness. From this point of view, I prefer the variable space in which bodies move in geodesics to a Euclidean space with a field of force. But I cannot regard the question as one concerning the facts.

The conclusion would seem to be, therefore, that, when physics is considered, as we are now considering it, as a deductive system, we do well to adopt the Einsteinian interpretation: free particles move in geodesics, and the law of gravitation is a law as to how geodesics are shaped in the neighbourhood of matter. This view is essentially simple, though it leads to complicated mathematics. It accords with the facts, and it puts the law of gravitation in a recognizable place among physical principles, instead of leaving it, as heretofore, an isolated and unrelated law. I propose, therefore, to continue to adopt Einstein's view as to the best way of interpreting the principles of physics, without suggesting that no other way is logically possible.

There is one matter of great theoretical importance, which is not very clear in the usual accounts of relativity. How do we know whether two events are to be regarded as happening to the same piece of matter? An electron or a proton is supposed to preserve its identity throughout time; but our fundamental continuum is a continuum of *events*. One must therefore suppose that one unit of matter is a series of events, or a series of sets of events. It is not clear what is the theoretical criterion for determining whether two events both belong to one such series. We may assume, I suppose, that two events which overlap—*i.e.* which are both present at some point of space-time—must belong to one unit of matter. (It is not to be assumed that an event which belongs to one unit of matter belongs to no other.) We may also assume that two events which have a space-like interval, or have a zero interval without overlapping, do not belong to one unit of matter. But when two events have a time-like interval, there is no obvious criterion. Any two such events can be connected by a geodesic in which any two points have a time-like separation; therefore, so far as the laws of dynamics are concerned, they *might* both belong to the same material unit. Yet sometimes we think they do, and sometimes we think they do not. It is evidently part of the business of physics to tell us how we are to decide this question in a given case. What can we say about it ?*

The decision must depend upon intermediate history—*i.e.* upon the existence of some series of intermediate events (or sets of events) following each other according to some law. If there exists any law which is in fact obeyed by strings of events, such a law can be used to define what we mean by one material unit. We know that there are such laws, but their importance in this connection is not emphasized, because it has hardly been realized that there is a problem owing to the substitution of events for bits of matter as the fundamental stuff of physics. For common sense, there is a more or less vague law of what may be called qualitative continuity. If you look persistently in a given direction, what you see, as a rule, alters gradually; there are exceptions, such as explosions, but they are rare. (I am not talking of a theoretical gradualness, but of one that is obvious to untrained perception.) If you see, say, a well-defined red patch, whose shape and tint do not alter greatly while you are looking, you conclude that there is a material object there, especially if you can touch it whenever you choose. Common sense achieves in this way a considerable measure of constancy in its objects. More is achieved by reducing matter to molecules, more still by reducing it to atoms, and yet more by reducing it to protons and electrons. But physicists would not feel pleased with

* This subject is considered again in Chap. XIV. from a somewhat different standpoint.

electrons and protons but for the fact that their tables and chairs, their laboratories and their books, consist, on the whole, of the same electrons and protons on different occasions. Qualitative continuity remains the basis of the whole proceeding. Suppose, one evening, you were to say to an astronomer: How do you know that that white patch in the sky is the moon? He would stare at you, and think you mad. He would *not* reply: because the course and phases of the moon have been worked out by astronomical theory, and that is where the moon ought to be, and the shape it ought to have, at the present moment in this latitude and longitude. What he would say is: Why, can't you *see* it's the moon? To which the right answer would be: Yes, *I* can, but I didn't suppose *you* could, because you ought to have got beyond such a crude criterion.

Moreover, there are identities in physics which are not material. A wave has a certain identity; if this were not the case, our visual perceptions would not have the intimate connection they in fact do have with physical objects. Suppose we see several lamps simultaneously: we are able to distinguish them because each sends out its own light-waves, which preserve their individuality until they reach the eye. Our chief reason for not regarding a wave as a physical object seems to be that it is not indestructible. But this is not our only reason, since, if it were, we might regard the energy of a wave as a physical object. We do not regard energy as a "thing," because it is not connected with the qualitative continuity of common-sense objects: it may appear as light or heat or sound or what not. But now that energy and mass have turned out to be identical, our refusal to regard energy as a "thing" should incline us to the view that what possesses mass need not be a "thing." We seem driven, therefore, to the view advocated by Eddington, that there are certain invariants, and that (with some degree of inaccuracy) our senses and our common sense have singled them out as deserving names. The correct theoretical definition of a single piece of matter will thus depend upon the mathematical invariants resulting from our formula for interval. This topic, however, demands a new chapter.

CHAPTER IX

INVARIANTS AND THEIR PHYSICAL INTERPRETATION

THERE is a point of view specially associated with Professor Eddington, which it is necessary to consider at this stage, since it arises naturally in the attempt to develop physics as a self-contained deductive system. According to this view, practically all theoretical physics is a vast tautology or convention, the only part excepted, so far, being the part which involves quantum-theory. This is not the whole of Professor Eddington's view on the subject, as he has shown when not writing simply as a technical physicist;* but it is what we may call his " professional " view.†

Let us begin with the conservation of momentum and of energy (or mass). Here we start from a proposition of pure mathematics. To explain this proposition will require certain preliminaries. It will be remembered that we had:

$$ds^2 = \Sigma g_{\mu\nu} dx_\mu dx_\nu.$$

We put:

$$g = \begin{vmatrix} g_{11} & g_{12} & g_{13} & g_{14} \\ g_{21} & g_{22} & g_{23} & g_{24} \\ g_{31} & g_{32} & g_{33} & g_{34} \\ g_{41} & g_{42} & g_{43} & g_{44} \end{vmatrix}$$

And we write $g^{\mu\nu}$ for the minor of $g_{\mu\nu}$ in this determinant, divided by g. Also:

$$g_\mu^\nu = \Sigma_\sigma g_{\mu\sigma} g^{\nu\sigma},$$

which $=0$ if $\mu \neq \nu$ and $=1$ if $\mu = \nu$.

The next step is the definition of the " three-index symbols," which are:

* See his essay in *Science, Religion, and Reality*, edited by Needham, 1925.

† Cf. *Mathematical Theory of Relativity*, §§ 52, 54, 66.

$$[\mu\nu,\sigma] = \tfrac{1}{2}\left(\frac{\partial g_{\mu\sigma}}{\partial x_\nu} + \frac{\partial g_{\nu\sigma}}{\partial x_\mu} - \frac{\partial g_{\mu\nu}}{\partial x_\sigma}\right)$$

$$\{\mu\nu,\sigma\} = \tfrac{1}{2}\sum_\lambda g^{\sigma\lambda}\left(\frac{\partial g_{\mu\lambda}}{\partial x_\nu} + \frac{\partial g_{\nu\lambda}}{\partial x_\mu} - \frac{\partial g_{\mu\nu}}{\partial x_\lambda}\right)$$

We can now define the tensor which Einstein uses for his law of gravitation. It is $G_{\mu\nu}$, where:

$$G_{\mu\nu} = \{\mu\sigma,a\}\ \{a\nu,\sigma\} - \{\mu\nu,a\}\ \{a\sigma,\sigma\} + \frac{\partial}{\partial x_\nu}\{\mu\sigma,\sigma\} - \frac{\partial}{\partial x_\sigma}\{\mu\nu,\sigma\}$$

summed for all values of σ and a from 1 to 4. Einstein takes as the law of gravitation $G_{\mu\nu}=0$ in empty space. For the moment, we are not concerned with the law of gravitation, but with certain identities. We put:

$$G = \Sigma g^{\mu\nu} G_{\mu\nu} \quad (\mu, \nu = 1, 2, 3, 4).$$

Further, there is a rule for raising or lowering suffixes in any tensor, of which an illustration is:

$$A^\mu_a = \sum_\nu g^{\mu\nu} A_{a\nu} \quad (\nu = 1, 2, 3, 4),$$

so that—

$$G^\nu_\mu = g^{\nu 1}G_{\mu 1} + g^{\nu 2}G_{\mu 2} + g^{\nu 3}G_{\mu 3} + g^{\nu 4}G_{\mu 4}.$$

Generalizing the notion of the "divergence" of a vector, we obtain a general definition of the divergence of any tensor. Taking a tensor of the form A^ν_μ for purposes of illustration, its "divergence" has four components:

$$(A^1_\mu)_1 + (A^2_\mu)_2 + (A^3_\mu)_3 + (A^4_\mu)_4 \quad (\mu = 1, 2, 3, 4),$$

where:

$$(A^1_\mu)_1 = \frac{\partial}{\partial x_1} A^1_\mu + \sum_a \{a1,1\}\, A^a_\mu - \sum_a \{\mu 1,a\}\, A^1_a,$$

and similarly for $(A^2_\mu)_2$, etc. These definitions have been given in order to enunciate the proposition:*

The divergence of $G^\nu_\mu - \tfrac{1}{2}g^\nu_\mu G$ *is identically zero,*

which Eddington calls "the fundamental theorem of mechanics."

* Eddington, *op. cit.*, p. 115.

In order to see the use made of this proposition, we need to introduce the "material energy-tensor," defined as:

$$T^{\mu\nu}=\rho_0\frac{\partial x_\mu}{\partial s}\frac{\partial x_\nu}{\partial s},$$

where ρ_0 is the "proper density" of the matter concerned—*i.e.* its density relative to axes moving with the matter. From this, by the usual rule for lowering a suffix, we obtain a tensor T^ν_μ. The principles of the conservation of mass and momentum are contained in the statement that the divergence of T^ν_μ vanishes. This suggests the identification of T^ν_μ with $G^\nu_\mu - \frac{1}{2}g^\nu_\mu G$, whose divergence vanishes identically—apart from a numerical factor, which, for convenience, is taken as -8π. Thus Eddington puts:

$$G^\nu_\mu - \tfrac{1}{2}g^\nu_\mu G = -8\pi T^\nu_\mu \quad\ldots\ldots\ldots\ldots\ldots (54\cdot 3),$$

which is the law of gravitation for continuous matter.

It has been necessary to make the above excursion into mathematical regions in order to be able to understand the observations which succeed to the above in Eddington's exposition (*op. cit.*, p. 119). He says:

"Appeal is now made to a Principle of Identification. Our deductive theory starts with the interval . . ., from which the tensor $g_{\mu\nu}$ is immediately obtained. By pure mathematics we derive other tensors. . . . These constitute our world-building material; and the aim of the deductive theory is to construct from this a world which functions in the same way as the known physical world. If we succeed, mass, momentum, stress, etc., must be the vulgar names for certain analytical quantities in the deductive theory; and it is this stage of naming the analytical tensors which is reached in (54·3). If the theory provides a tensor $G^\nu_\mu - \frac{1}{2}g^\nu_\mu G$, which behaves in exactly the same way as the tensor summarizing the mass, momentum and stress of matter is observed to behave, it is difficult to see how anything more could be required of it."

There are a number of other examples of the same method in Eddington's work, but we may take the above as typical, since it is the simplest mathematically. It is worth while to consider the nature of the method, apart from its technical embodiment. This is the more necessary, as it is not easy to be clear as to the logical and empirical elements in theoretical physics as developed by the above method.

Fundamentally, the method is the same as that which has always been pursued when mathematics has been applied to the physical world. The aim has been to obtain mathematical laws which gave correct results wherever they could be tested by observation. The fewer and more general and more comprehensive the laws, the more scientific taste was gratified. Newton's law of gravitation was better than Kepler's laws, both because it was one law instead of three, and because it gave a larger number of correct deductions. But at every stage the subject-matter of physics grows more abstract, and its connection with what we observe grows more remote. Eddington's ideal is to start with only one fundamental law —namely, the formula for ds^2—which, as generalized by Weyl, will give electromagnetic equations as well as gravitation. From this one fundamental law, by pure mathematics, we deduce the existence of quantities behaving in certain ways. Elementary theorizing from observation has led us to believe that there are quantities connected with what we observe which behave in these ways. We therefore identify the observed quantities with the deduced quantities. This is, in essence, the same sort of thing as we do when we associate what we see with light-waves. We may thus regard physics from the two points of view, the inductive and the deductive. In the latter, we start from the formula for interval (together with certain other assumptions), and we deduce by mathematics a world having certain mathematical characteristics. In the inductive view, the same mathematical characteristics are arrived at, but they are now those which may be supposed to belong to the physical world in its entirety if we supplement observation by means of the postulate that everything happens in accordance with simple general laws.

We may thus say that the world of elementary physics is semi-abstract, while that of deductive relativity-theory is wholly abstract. The appearance of deducing actual phenomena from mathematics is delusive; what really happens is that the phenomena afford inductive verification of the general principles from which our mathematics starts. Every observed fact retains its full evidential value; but now it confirms not merely some particular law, but the general law from which the deductive system starts. There is, however, no logical necessity for one fact to follow given another, or a number of others, because there is no logical necessity about our fundamental principles.

The question of interpretation, it must be admitted, is somewhat difficult when physics is conceived in this very abstract manner. What, for example, is ds? We start from a view which is, to a certain extent, intelligible in terms of observation. In the case of a time-like interval, it is the time which elapses between the two events according to a clock, not subject to constraints, which is present at both events. On the earth's surface, the time measured by a clock can be inferred, with suitable precautions, from the visual perceptions of a careful observer. In the case of a space-like interval, ds is the distance between two events as estimated by measurements carried out on a body which is present at both, and for which the two events are simultaneous. The elementary operation of measuring lengths is here supposed possible. But when we pass from this initial view to the abstract view which is required by the general theory of relativity, the interval can only be actually estimated by using rather elaborate physics to make deductions from what can be actually observed by means of clocks and footrules. For logical theory, the interval is primitive, but from the point of view of empirical verification it is a complicated function of empirical data, deduced by means of physics in its semi-abstract form. The unity and simplicity of the deductive edifice, therefore, must not blind us to the complexity of empirical physics, or to the logical independence of its various portions.

In particular, when the conservation of mass or of momentum appears as an identity, that is only true in the deductive system; in their empirical meaning, these laws are by no means logical necessities. There might easily be a world in which they were false, and it might be capable of a treatment as unified and mathematical as the general theory of relativity; but, if so, the fundamental laws would be different.

What is novel and interesting in the point of view we have been considering is the character of the relation between empirical and deductive physics. But there is no real diminution of the need for empirical observation. I do not for a moment suggest that anything in the above is a criticism of Professor Eddington; indeed, I imagine he would regard it as a string of truisms. I have been concerned only to guard against a possible misunderstanding on the part of those who do not feel for mathematics the contempt which is bred of familiarity.

In the foregoing remarks, however, we have neglected one important aspect of Eddington's theory. In addition to the fact that the whole general theory of relativity can be deduced from a few simple assumptions, interest attaches to the manner of the deduction and the considerations by which the substantial import of mathematical formulæ is made less, or at least other, than would naturally be supposed. A good example is afforded by a paragraph headed "Interpretation of Einstein's Law of Gravitation."* The law concerned is not

* *Op. cit.*, § 66, pp. 152-155.

$G_{\mu\nu}=0$, which is not supposed to be quite accurate where stellar distances are concerned; it is the modified law:

$$G_{\mu\nu}=\lambda g_{\mu\nu},$$

where λ must be very small, so small that within the solar system the new law gives the same results, within the limits of observation, as $G_{\mu\nu}=0$. The new law is shown to be equivalent to the assumption that, in empty space, the radius of curvature in every direction is everywhere $\sqrt{3/\lambda}$. But this is interpreted as a law about our measuring rods—namely, that they adjust themselves to the radius of curvature at any place and in any direction. It is interpreted as meaning:

"The length of a specified material structure bears a constant ratio to the radius of curvature of the world at the place and in the direction in which it lies." And the following gloss is added:

"The law no longer appears to have any reference to the constitution of an empty continuum. It is a law of material structure showing what dimensions a specified collection of molecules must take up in order to adjust itself to equilibrium with the surrounding conditions of the world."

In particular, electrons must make these adjustments, and it is suggested elsewhere that the symmetry of an electron and its equality with other electrons are not substantial facts, but consequences of the method of measurement (pp. 153-4). One cannot complain of an author for not doing everything, but at this point most readers will feel a desire for some discussion of the theory of measurement. The elementary meaning of measurement of lengths is derived from superposition of a supposedly rigid body. A rigid body, as Dr Whitehead has pointed out, is primarily one which *seems* rigid, such as a steel bar in contradistinction to a piece of putty. When I say that a body "seems" rigid, I mean that it looks and feels as if it were not altering its shape and size.

This, so far as it can be relied upon, implies some constant relation to the human body: if the eye and the hand grew at the same rate as the "rigid" body, it would look and feel as if it were unchanging. But if other objects in our immediate environment did not grow meanwhile, we should infer that we and our measure had grown. There would, however, be no meaning in the supposition that all bodies are bigger in certain places than they are in certain others; at least, if we suppose the alteration to be in a fixed ratio. If we do not add this proviso, there is a good meaning in the supposition; in fact, we do actually believe that all bodies are bigger at the equator than at the North Pole, except such as are too small to be visible or palpable. When we say that the length of an object at the equator is one metre, we do not mean that its length is that which the standard metre would have if moved from Paris to the equator. But the expansion of bodies with temperature would have been difficult to discover if it had not been possible to bring bodies of different temperatures into the same neighbourhood and measure them before their temperatures had become equal; it would also have been difficult if all bodies had expanded equally when their temperatures rose. These elementary considerations, along with many others, make rigidity an ideal, which actual bodies approach without attaining. Mere superposition thus ceases to give a measure of length: it gives still a comparison of the two bodies concerned, but not of either with the standard unit of length. To obtain the latter, we have to adjust the immediate results of the operation of measuring, by means of a mass of physical theory. If the measures which we obtain are mutually consistent, that is all we can ask; but it is possible that a change in physical theory might have given other measures which would also have been mutually consistent.

Professor Eddington, in the passage which we quoted partially in introducing this discussion, is careful to say that he is concerned with measurement by direct comparison. He says:

"The statement that the radius of curvature is a constant length requires more consideration before its full significance is appreciated. Length is not absolute, and the result can only mean *constant relative to the material standards of length* used in all our measurements and in particular in those measurements which verify $G_{\mu\nu}=\lambda g_{\mu\nu}$. In order to make a direct comparison the material unit must be conveyed to the place and pointed in the direction of the length to be measured. It is true that we often use indirect methods, avoiding actual transfer or orientation; but the justification of these indirect methods is that they give the same result as a direct comparison, and their validity depends upon the truth of the fundamental laws of nature. We are here discussing the most fundamental of these laws, and to admit the validity of the indirect methods of comparisons at this stage would land us in a vicious circle."

I confess that I am puzzled by this passage. Taken in its plain and obvious sense, it means that the standard metre is to be taken from Paris, and used without any corrections for temperature, etc., because as soon as we introduce such corrections we are assuming a great deal of physics, and thus seem to be making ourselves liable to the vicious circle which, we are told, is to be avoided. It is evident, however, that this is not what Professor Eddington means, since he goes on at once to speak of the electron as making the adjustments concerned. Now the electron may be, theoretically, a perfect spatial unit, but we certainly cannot compare its size with that of larger bodies *directly*, without assuming any previous physical knowledge. It seems that Professor Eddington is postulating an ideal observer, who can see electrons just as directly as (or, rather, much more directly than) we can see a metre rod. In short, his "direct measurement" is an operation as abstract and theoretical as his mathematical symbolism.

That being admitted, we may take the electron as our spatial unit, and ask ourselves what our ideal observer could do with it. He could not take a lot of electrons and place them end on in a row, with a view to measuring a given length, since an infinite force is required to make two electrons touch. To measure ordinary lengths, he would have to take (say) hydrogen at a given temperature and pressure, enclosed in a balloon whose radius is the length to be measured; he could then count the number of electrons in the balloon and take its cube root as a measure of the said length. But to ascertain the temperature and pressure, he will have to make other measurements; moreover, he will have to *assume* that his balloon is spherical. Altogether, the method does not seem very practical.

I have no complete theory of physical measurements to offer, but it seemed desirable to illustrate how difficult it is to say precisely what measurement means in an advanced science such as physics. We have certain postulates, such as "lengths which are equal to the same length are equal to one another," but actual measurements, when made with sufficient accuracy, are not found to verify these postulates. Therefore we invent physical laws to save the postulates. With each fresh law it becomes more difficult to say exactly what we do mean when, *e.g.*, we give the wave-length of a certain line in the spectrum of hydrogen in terms of the metre. (This is particularly odd in view of the fact that these wave-lengths are given to more significant figures than can be warranted by the operations applicable to the standard metre itself, whose length is only known, in comparison with other lengths, to a very moderate degree of approximation.) In physical theory, measurement should rest upon an integration of the formula for ds^2. But in physical practice the $g_{\mu\nu}$ of that formula can only be determined by means of measurements. Thus the only thing we seem warranted in saying is this: It is possible to correct the results of actual measurements according to certain known rules, in such a way that the corrected lengths shall satisfy such postulates as Euclid's first axiom; when this is done, we find, by means of physical theory, that all electrons have the same size. But this is not, considered empirically, at all a simple fact. And considered as a statement in the deductive theory it probably has a good meaning, but one which demands much elucidation. Until this is forthcoming, all use of numbers as measures of physical quantities in theoretical physics raises problems, since we do not know what, in theoretical physics, replaces the operation of measurement as conducted in the laboratory and in daily life.

The theory of length-measurement raises problems which bring us naturally to Weyl's relativistic theory of electromagnetism, which we must now briefly consider.

CHAPTER X

WEYL'S THEORY

THE theory to be considered in this chapter is, from a geometrical point of view, a natural generalization of Einstein's arbitrariness of co-ordinates; from a physical point of view, it fits electromagnetism into the deductive system, which Einstein's theory does not do. The theory is due to Hermann Weyl, and will be found in his *Space, Time, Matter* (1922).

The puzzles about measurement considered at the end of Chapter IX. naturally suggest the point of view from which Weyl starts. As he says: " The same certainty that characterizes the relativity of motion accompanies the principle of the *relativity of magnitude* " (*op. cit.*, p. 283). Measurement is a comparison of lengths, and Weyl suggests that, when lengths in different places are to be compared, the result may depend upon the route pursued in passing from the one place to the other. Lengths at the same place (*i.e.* having one end identical), if small, he regards as directly comparable; also he assumes continuity in the changes accompanying transportation. This is not the sum-total of his assumptions, nor the most general way of stating them; but before we can state them adequately certain explanations are necessary.

Reduced to its simplest terms, the conception used by Weyl may be expressed as follows. Given a vector at a point, what are we to mean by the statement that a vector at another point is equal to it? There must be some element of convention in our definition; let us therefore, as a first step, set up a unit of length in each place, and see what limitations it is desirable to impose on our initial arbitrariness.

There is, to begin with, an assumption which is made almost

tacitly, and that is, that we can recognize something in one place as the "same" vector as something at another place. We may perhaps take this sameness as being merely analytical: the two are the same function of the co-ordinates at their respective places. I do not think this is all that is meant, since a vector is supposed to have some physical significance; but if more is meant, it is not clear how it is to be defined. We will therefore assume that, given a function of the co-ordinates which is a vector, we shall regard the same function of other values of the co-ordinates as the "same" vector at another place.

We next have to define "parallel displacement." This may be defined in various ways. Perhaps the most graphic description is to say that it is displacement along a geodesic (Eddington, *op. cit.*, p. 71). Another definition is that it is a displacement such that the "covariant derivative" vanishes, the covariant derivative of a vector A_μ with respect to ν being defined as $A_{\mu\nu}$, where:

$$A_{\mu\nu} = \frac{\partial A_\mu}{\partial x_\nu} \quad \sum_{a} \{\mu\nu, a\} A_\alpha \quad (\alpha = 1, 2, 3, 4).$$

For the definition of $\{\mu\nu, a\}$, see the beginning of Chapter IX. In the tensor calculus, covariant differentiation takes the place of ordinary differentiation for many purposes, since the covariant derivative of a tensor is a tensor, whereas the ordinary derivative is in general not a tensor. We assume that our units of length in different places are so chosen that, when a small displacement is moved to a neighbouring place by parallel displacement, the change in the measure of its length is small, and is proportional to its length. We assume, in short, that the ratio of the increase of length to the initial length for a change of co-ordinates (dx_1, dx_2, dx_3, dx_4) is:

$$\kappa_1 dx_1 + \kappa_2 dx_2 + \kappa_3 dx_3 + \kappa_4 dx_4.$$

So that $(\kappa_1, \kappa_2, \kappa_3, \kappa_4)$ form a vector, κ_μ.

Now it is possible to express Maxwell's equations in terms of a vector which may be identified with the above vector κ_μ. Hence it is possible to regard electromagnetic phenomena as explained by the variation of what is taken as the unit as we pass from point to point. I shall not attempt to explain the theory, as it would in any case be necessary to read a full account in order to grasp its significance.

Here, perhaps even more than elsewhere in relativity theory, it is difficult to disentangle the conventional elements from those having physical significance. On the face of it, it might seem as though we were attempting to account for actual physical phenomena by means of a mere convention as to choice of units. But this, of course, is not what is meant. The way the unit is assigned in different places is called by Eddington the "gauge-system": this is only partially arbitrary, and is in part the representation of the physical state of the world. This has to do with the fact that vectors are not purely analytical expressions, but also correspond to physical facts. It would seem, however, that the theory has not yet been expressed with the logical purity that is to be desired, chiefly because it is not prefaced by any clear account of what is to be understood by "measurement"—or, what comes to much the same thing from the standpoint of theory, what we are to mean when we talk of "moving" a vector, whether by parallel displacement or in any other way. To "move" something, we must be able to recognize some identity between things in different places. Perhaps all this is quite clear in the minds of competent exponents of the theory, but if so they have not succeeded in conveying their thoughts without loss of clarity to readers who have not their background. When Eddington says: "Take a displacement at P and transfer it by parallel displacement to an infinitely near point P'" (p. 200), I find myself wondering how, exactly, the displacement is to preserve its identity throughout the transfer, and the only answer suggested by the accompanying formulæ is that the identity is that of an algebraic expression in terms of the co-ordinates. This, however, is clearly insufficient.

Professor Eddington, after expounding Weyl's theory, proceeds to generalize it, and some of his accompanying elucidations are relevant to our present difficulties. Thus he says (p. 217):

"In Weyl's theory, a gauge-system is partly physical and partly conventional; lengths in different directions but at the same point are supposed to be compared by experimental (optical) methods; but lengths at different points are not supposed to be comparable by physical methods (transfer of clocks and rods), and the unit of length at each point is laid down by a convention. I think this hybrid definition of length is undesirable, and that length should be treated as a purely conventional or else a purely physical conception."

He proceeds to a generalized theory in which, at first, length is *purely* conventional, for comparisons at a point as well as for comparisons between different points. This generalized theory does not seem to involve the same kind of difficulties as those which have been troubling us. The following passage, for example, states the matter with great clearness (p. 226):

"The relation of displacement between point-events and the relation of 'equivalence' between displacements form part of one idea, which are only separated for convenience of mathematical manipulation. That the relation of displacement between A and B amounts to such-and-such a quantity conveys no absolute meaning; but that the relation of displacement between A and B is equivalent to the relation of displacement between C and D is (or at any rate may be) an absolute assertion. Thus four points is the minimum number for which an assertion of absolute structural relation can be made. The ultimate elements of structure are thus four-point elements. By adopting the condition of affine geometry, I have limited the possible assertion with regard to a four-point element to the statement that the four points do, or do not, form a parallelogram. The defence of affine geometry thus rests on the not unplausible view that four-point elements are recognized to be differentiated from one another by a single character—viz. that they are or are not of a particular kind which is conventionally named parallelogramical. Then the analysis of the parallelogram property into a double equivalence of *AB* to *CD* and *AC* to *BD*, is merely a definition of what is meant by the equivalence of displacements."

Here we have a logically satisfactory theoretical basis for a metric. We may suppose that, as a matter of fact, there are important properties of groups of four points which are " parallelogramical," and that actual physical measurement is an approximate method of discovering which groups have this property. We shall find certain laws approximately fulfilled by rough-and-ready measurements, and fulfilled with increasing accuracy as we introduce refinements into the process of measurement. Consider, for example, Euclid's first axiom: Things which are equal to the same thing are equal to one another. Presumably Euclid regarded this as a logically necessary proposition, and so do people who are engaged in the practice of measurement. If two lengths each equal to a metre are found to be not equal to each other, the plain man assumes that there must be a mistake somewhere. We are therefore continually redefining the actual operations of measurement with a view to verifying Euclid's first axiom as nearly as possible. But with the above-quoted definition of equality of length the first axiom becomes a substantial proposition, namely: If *ABCD* is a parallelogram, and likewise *DCEF*, then *ABEF* is a parallelogram. If this proposition is true, then it is theoretically possible to define measurement in such a way that two lengths each equal to a metre shall always be equal to each other. What is called " accuracy " is, speaking generally, an attempt to obtain a result conformable with some ideal standard supposed to be logical but in fact physical.

What do we mean by saying that a length has been "wrongly" measured? Whatever result we obtain from measuring a given length, the result represents a fact in the world. But in what we call a "wrong" measurement, the fact ascertained is complex and of small universality. If the observer has simply misread a scale, the fact ascertained involves reference to his psychology. If he has neglected a physical correction—*e.g.* for the temperature of his measure—the fact refers only to a measurement carried out with that particular apparatus on that particular occasion. In relativity theory we have another set of what might be called "inaccurate" measurements—*e.g.* measurements of the masses of α-particles or β-particles emitted from radio-active bodies must be corrected for their motion relative to the observer before they acquire any general significance. It is always the search for simple relations which enter into general laws that governs successive refinements. But the existence of such relations (where they do exist) is an empirical fact, so that much that seems *prima facie* to be logically necessary is really contingent. On the other hand, the number of premisses in a deductive system which has to agree with an empirical science can, by logical skill, be diminished to an extent which may be astonishing. Of this, the theory of relativity is a very remarkable example. The theory is a combination of two diverse elements: on the one hand, new experimental data; on the other, a new logical method. It must be regarded as a happy accident that the two appeared together; if the right kind of theoretical genius had not happened to be forthcoming, we might have had to be content for a long time with patched-up hypotheses such as the FitzGerald contraction. As it is, the combination of experiment and theory has produced one of the supreme triumphs of human genius.

CHAPTER XI

THE PRINCIPLE OF DIFFERENTIAL LAWS

THROUGHOUT the theory of relativity, there is an application, with increasing stringency, of a principle which begins to make itself felt in physics with Galileo, in spite of the fact that he did not possess the mathematical technique which it demands. The principle I mean is that of " differential laws," as it may be called. This means that any connection which may exist between distant events is the result of integration from a law giving a rate of change at every point of some route from the one to the other. One may give a simple illustration of a differential law from the " curve of pursuit ": a man is walking along a straight road, and his dog is in a field beside the road; the man whistles to the dog, and the dog runs towards him. We suppose that at each moment the dog runs exactly towards where his master is at that moment. To discover the curve described by the dog is a problem in integration, which becomes definite given certain further data. The Newtonian law of gravitation gives a very similar type of law, except that it is the acceleration of the planet, not its velocity, that is directed towards the sun at each moment. It has long been a commonplace of physics that its causal laws should have this differential character: they should tell primarily a tendency at each moment, not the outcome after a finite time. In a word, its causal laws take the form of differential equations, usually of the second order.

This view of causal laws is absent from quantum theory, from the ideas of savages and uneducated persons, and from the works of philosophers, including Bergson and J. S. Mill. In quantum theory, we have a discrete series of possible sudden changes, and a certain statistical knowledge of the proportion of cases in which each possibility is realized; but we have no knowledge as to what determines the occurrence of a particular change in a particular case. Moreover, the change is not of the sort that can be expressed by differential equations: it is a change from a state expressed by one integer or set of integers to a state expressed by another. This kind of change may turn out to be physically ultimate, and to mark out at least a part of physics as governed by laws of a new sort. But we are not likely to find science returning to the crude form of causality believed in by Fijians and philosophers, of which the type is "lightning causes thunder." It can never be a law that, given *A* at one time, there is sure to be *B* at another time, because something might intervene to prevent *B*. We do not derive such laws from quantum phenomena, because we do not, in their case, know that *A* will not continue throughout the time in question. The natural view to take at present is that quantum phenomena have to do with the interchange of energy between matter and the surrounding medium, while continuous change is found in all processes which involve no such interchange. There are, however, difficulties in any view at present, and it is not for a layman to venture an opinion. It seems not improbable that, as Heisenberg suggests, our views of space-time may have to be modified profoundly before harmony is achieved between quantum phenomena and the laws of transmission of light *in vacuo*. For the moment, however, I wish to confine myself to the standpoint of relativity theory.

Although physics has worked with differential equations ever since the invention of the calculus, geometry was supposed to be able to start with laws applying to finite spaces. If we accept the Einsteinian point of view, there can no longer be any separation between geometry and physics; every proposition of geometry will be to some extent causal. Take first the special theory. Relatively to axes (x, y, z, t) we can obtain propositions of geometry by keeping t constant; but relatively to other axes these propositions will refer to events at different times. It is true that these events, in any system of co-ordinates, will have a space-like interval, and will have no direct causal relations with each other; but they will have indirect causal relations derived from a common ancestry. Let us take some example, say: The sum of the angles of a triangle is two right angles. Our triangle may be composed of rods or of light-rays. In either case, it must preserve a certain constancy while we measure it. Both rods and light-rays are complicated physical structures, and the physical laws of their behaviour are involved in taking them as approximations to ideal straight lines. Nevertheless, so far as the special theory is concerned, all this might be allowed, and yet we might maintain a certain distinction between geometry and physics, the former being a set of laws supposed exact, and approximately verified, for the relations of the x, y, z co-ordinates in any Galilean frame when t is kept constant.

But in the general theory the intermixture of geometry and physics is more intimate. We cannot accurately reduce ds^2 to the form:

$$dx_4^2 - dx_1^2 - dx_2^2 - dx_3^2,$$

and therefore we cannot accurately distinguish one co-ordinate as representing the time. We cannot therefore obtain a timeless geometry by putting x_4=constant. With this goes a change in our axioms. We no longer have, as in Euclid, in Lobatchevsky and Bolyai, and in projective geometry, axioms dealing with straight lines of finite length. We have now only, as our initial apparatus, a geometry of the infinitesimal, from which large-scale results must be obtained by integration. From this point of view, Weyl's extension of Einstein appears natural. As we saw in the last chapter, quoting Eddington, the statement that the distances AB,

CD are equal is the assertion of a relation between the four points, *A*, *B*, *C*, *D*. If all the relations which constitute our initial apparatus are to be confined to the infinitesimal, so must this relation; if so, *A*, *B*, *C*, *D* must all be close together, and Weyl's geometry results.

At this point, however, the pure mathematician is likely to feel a difficulty which does not greatly trouble the physicist. The physicist thinks of his infinitesimals as actual small quantities, which may—*e.g.* in astronomical problems—be such as would be reckoned large in other problems. For him, therefore, a statement in terms of infinitesimals is quite satisfactory. But for the pure mathematician there are no infinitesimals, and all statements in which they seem to occur must be expressible as limits of what happens to finite quantities. To take our particular case: We must be able to say of a small finite quadrilateral that it is *approximately* a parallelogram, if we are to be able to assign a meaning to the statement that an infinitesimal quadrilateral may be *accurately* a parallelogram. The case is exactly analogous to velocity in elementary kinematics: we can assign a meaning to velocity only because we can measure finite distances and times, and so form the conception of the limit of their quotient. It is not wholly clear how we are to satisfy this requirement in the case of Weyl's theory. I think, however, that there is not the slightest reason to suppose that it cannot be satisfied. Let "*R* (*a*, *b*, *c*, *d*)" mean "*a*, *b*, *c*, *d* form a parallelogram." We are supposed to have also *R*(*adcb*), *R*(*badc*), etc., but not *R*(*acbd*), etc. Also if we have *R*(*abcd*) and *R*(*cdef*), we are to have *R*(*abfe*). But if we take "*S* (*a*, *b*, *c*, *d*)" to mean "*a*, *b*, *c*, *d* form an approximate parallelogram," we cannot (if there

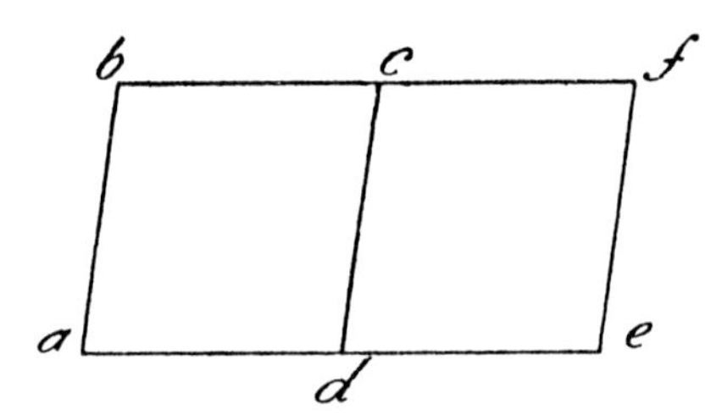

is any way of specifying a degree of approximation) argue from $S(abcd)$ and $S(cdef)$ to $S(abfe)$. Now if we assume, as Weyl does, that lengths at a given point are comparable, we can perhaps give the necessary definitions. We shall have to take S, not R, as our fundamental relation, since the distance between any two points is finite, and it is assumed that no finite quadrilateral can be accurately a parallelogram. Or perhaps we shall have to go a step further, and take as fundamental a relation of eight points, say

$$(abcd)T(efgh),$$

meaning " *abcd* is more nearly a parallelogram than *efgh*." We shall then say that, given any four points, a, b, e, f, it is possible to find points c, d nearer to b and a respectively than e and f are, such that

$$(abcd)T(abfe).$$

Further, we can say that, if a, b, c, d are sufficiently near together, and

$$(abcd')T(abcd),$$

then the ratio of dd' to dc can be made to approach zero as a limit by diminishing the size of *abcd* in a purely ordinal sense. (Ordinal relations among points, as we saw earlier, are presupposed in the theory of relativity.)

It is highly probable that the above process can be simplified. It is, however, of no importance in itself; its only purpose is to show that the derivatives required can be correctly defined, and that, however the mathematical treatment may confine itself to infinitesimals, relations between points whose distances are finite must be presupposed if the infinitesimal calculus is to be applicable.

This last result, whose generality is obvious from the theory of limits, is of some philosophical importance. Wherever mathematics works in a continuous medium with relations which may be loosely described as next-to-next, there must be other relations, holding between points at finite distances from each other, and having the next-to-next relations as their limits. Thus, when we say that laws have to be expressed by differential equations, we are saying that the finite relations which occur cannot be brought under accurate laws, but only their limits as distances are diminished. We are not saying that these limits are the physical realities; on the contrary, the physical realities continue to be the finite relations. And if our theory is to be adequate, some way must be found of so defining the finite relations as to make the passage to the limit possible.

It is considered a merit in the general theory of relativity, particularly in Weyl's form (or the still more general form suggested by Eddington), that it dispenses with what we may call "integrated" relations as regards its fundamentals. Thus Eddington, after pointing out that he is concerned with structure, not with substance, proceeds (p. 224):

"But structure can be described to some extent; and when reduced to ultimate terms it seems to resolve itself into a complex of relations. And further these relations cannot be entirely devoid of comparability; for if nothing in the world is comparable with anything else, all parts of it are alike in their unlikeness, and there cannot be even the rudiments of a structure.

"The axiom of parallel displacement is the expression of this comparability, and the comparability postulated seems to be almost the minimum conceivable. Only relations which are close together—*i.e.* interlocked in the relation-structure—are supposed to be comparable, and the conception of equivalence is applied to only one type of relation. This comparable relation is called displacement. By representing this relation graphically we obtain the idea of location in space; the reason why it is natural for us to represent this particular relation graphically does not fall within the scope of physics.

"Thus our axiom of parallel displacement is the geometrical garb of a principle which may be called 'the comparability of proximate relations.'"

It is obvious that, in the above passage, Eddington is *imagining* displacements at a small finite distance from each other, not at an infinitesimal distance; he is not thinking of all the apparatus involved in a procedure which replaces infinitesimals by limits. One might suggest that he is supposing, *e.g.*, that a footrule will not change much during the portion of a second required to transfer it from one part of a given page to another. But when we say that it will not change "much," we imply some standard of quantitative comparison other than the footrule; and this leads to the problems we have been considering.

I cannot but think that Eddington's point of view lends itself to development and further analysis by means of mathematical logic; in particular, this applies to the conditions for the possibility of measurement, a subject which will be considered explicitly in the next chapter. But for the present my concern is with "the comparability of proximate relations." In the first place, what is meant by "comparability"? A moment's reflection shows that what is wanted is a symmetrical transitive relation which each of the relations in question has to some others, but not to all. (It is assumed, in the particular case of Eddington's general geometry, that when there is such a relation of the interval *ab* to the interval *cd*, there is also such a relation of the interval *ad* to the interval *bc*. But this, as he admits (p. 226), is not essential.) Now why should we suppose that a transitive symmetrical relation of the above sort is more likely to exist between small intervals than between large ones? *I.e.*, if b' is between *a* and *b*, and c' between *d* and *c*, is it more likely that the relation in question will hold between ab' and dc' than between *ab* and *dc*? I do not see why we should think so. And I think further that, with a correct interpretation of infinitesimals, the whole belief that causation must always be from next-to-next becomes untenable unless continuity is abandoned. Causal laws may all be differential equations, but the grounds for thinking that they are must be empirical, not *a priori*. They cannot be derived from the impossibility of action at a distance unless distance itself is a derivative from causality, which may well be the case, but does not represent any part of the views of those who are anxious to dispense with action at a distance. It may well be, therefore, that there is one department of physics—that included in the general theory of relativity, as supplemented by Weyl—in which everything proceeds by differential equations, while there is another part—that dealt with by quantum theory—in which this whole apparatus is inapplicable. There is absolutely no *a priori* reason why everything should go by differential equations, since, even then, causation does not really go from next-to-next: in a continuum there is no "next." It is, at bottom, because "next-to-next" seems natural that we like a procedure of differential equations; but the two are logically incompatible, and our preference for the second on account of the first proceeds only from logical confusion.

CHAPTER XII

MEASUREMENT

REPEATEDLY, in previous discussions, we have come up against the problem of measurement. It is time to consider it on its own account, both how it is to be defined, and in what circumstances it is possible.

In the first place, what do we mean by measurement? Clearly we do *not* mean *any* method of assigning numbers to a collection of objects; there must be properties of importance connected with the numbers assigned. We do not say that the books in the British Museum are "measured" by their press-marks. Given any collection whose cardinal number is less than or equal to $2^{\aleph_0}$, we can assign some or all of the real numbers as "press-marks" of the several members of the collection. Given any collection of $2^{\aleph_0}$ terms, it can be arranged in a Euclidean or non-Euclidean space of any known sort with any finite number of dimensions, and when so arranged it will be amenable to the whole of metrical geometry. But the "distance" between two terms of the collection, when it is defined in this way, will, in general, be quite unimportant, in the sense that it will have only such properties as follow tautologically from its definition, not such further empirical properties as would make the definition valuable. So long as this is the case, there is no reason to prefer one to another of the various incompatible systems of distances which pure mathematics would allow us to assign.

Let us take an illustration. In projective geometry we start from a set of axioms which say nothing about quantity, and do not even *obviously* involve order. But it is found that they do lead to an order, and that, by means of the order, co-ordinates can be assigned to points. These co-ordinates have a definite projective meaning: they represent the series of quadrilateral constructions required to reach the point in question from certain given initial points. (I omit complications concerning limits; these are dealt with in the chapter "Projective Geometry" in *The Principles of Mathematics.*) In this case, it may seem doubtful whether we have measurement or not. We have assigned co-ordinates in a manner which preserves the order-relations of points, and it turns out that the ordinary distance between two points is a simple function of their projective co-ordinates, though the function is somewhat different according as space is Euclidean, hyperbolic, or elliptic. It is just because of this difference that we shall *not* say we have "measured" distances when we have introduced projective co-ordinates. These co-ordinates, for example, will not tell us, even approximately, how long it would take to walk from one place to another, and this is the sort of thing that measurement ought to tell us.

What, then, *is* meant when it is said that, in the theory of relativity, there is a metrical relation of interval? Let us take up the matter at the point where Eddington leaves it. He suggests that all that is needed is "comparability" between two point-pairs, or, as he says, between two "displacements." (We may leave aside for the moment the question whether this is only to hold for point-pairs which are very near together.) This language seems somewhat vague; let us try to give it precision.

Suppose that between two point-pairs there is sometimes, but not always, a symmetrical transitive relation S. Then we can define as "the distance between x and y" the class of all point-pairs having the relation S to (xy). If now instead of $(xy)S(zw)$ we write $xy=zw$, we shall have:

If $xy=zw$, then $zw=xy$;
If $xy=zw$ and $zw=uv$, then $xy=uv$.

From these two it follows that every pair of objects x, y in the field of S is such that

$$xy = xy.$$

This seems to be as much as is strictly implied by Eddington's words, but it is certainly not all that we need. Nor does it become sufficient if we add:

$$\text{If } xy = zw, \text{ then } xz = yw.$$

There must be a connection between distances and ordinal relations, there must be ways of adding distances, and there must be ways of inferring new distances from a certain number of data, as in $ds^2 = \Sigma g_{\mu\nu} dx_\mu dx_\nu$. If all these conditions are fulfilled, we can then proceed to ask whether our distances have any further important physical properties.

The sort of relation that will not do is illustrated if we take $xy = zw$ to mean that xy and zw have the same apparent dimensions in the visual field of a certain observer—*e.g.* the diameters of the sun and moon will approximately have this relation, which is symmetrical and transitive, but physically unimportant. Let us see what is necessary in order to get a definition of distance which will have as many as possible of the properties possessed by distance in elementary geometry.

If we confine ourselves to three dimensions, we can at once define a plane: it will consist of all points equidistant from two given points. The points in this plane which are equidistant from two given points in it lie on a straight line; we may take this as the definition of a straight line. Thus given two points, P, Q, we can define the middle point M of PQ; it is the point on PQ which is equidistant from P and Q. We shall need an axiom to the effect that this point always exists and is always unique. Thus we can halve distances and double them: we shall of course define PM as half of PQ. From this point onwards, the assignment of numerical measures to our distances offers no difficulty. It is therefore only necessary to scrutinize what has already been said.

In ordinary Euclidean geometry, there is exactly one point on a plane which is equidistant from three given points on the plane; it is the centre of the circumscribed circle. In three dimensions, there is one point equidistant from four given points; in four, from five. This last holds also in the special theory of relativity, and even in the general theory so long as the distances concerned are small. If we take a point $(d_1x_1, d_1x_2, d_1x_3, d_1x_4)$ near the origin, another point (dx_1, dx_2, dx_3, dx_4) is equidistant from this point and the origin if $\Sigma g_{\mu\nu} dx_\mu d_1x_\nu = d_1s^2$ (where the $g_{\mu\nu}$ have their values at the origin), which is a simple equation in dx_μ. Four such equations give a unique set of values for (dx_1, dx_2, dx_3, dx_4). Thus there is just one point equidistant from five given points close together. Moreover, a simple equation, which we may take to be that of the part of a plane near the origin, gives the locus of points near the origin and equidistant from it and a neighbouring point. In fact, as we should expect, for small distances everything proceeds as in elementary geometry, given the formula for ds^2.

But the mere assumption that there is such a relation as S between point-pairs does not yield these results, since it does not imply the interrelation of distances which is given by the formula for ds^2. Nevertheless, it does suffice theoretically as a basis of measurement, since, as we have seen, it enables us to halve distances and double them, and therefore to assign numbers to them. This shows that the geometry of relativity, even in its most general and abstract form, assumes a good deal more than the mere possibility of measurement, which, in itself, is of very little value. In itself, it does not lead to a geometry; this only results when there is some interconnection between different measures.

It may be asked whether, when the geometry of relativity is generalized to the utmost, any genuinely quantitative element remains in its formulæ. We start with an ordered four-dimensional manifold, and we assign co-ordinates subject to the sole restriction that their order-relations are to reproduce those of the given manifold. We then proceed to find formulæ (tensor-equations) which hold equally in all systems of co-ordinates satisfying the above condition. It might seem a possibility that such formulæ really express only ordinal relations, and that the sole advantage of co-ordinates lies in the fact that they provide names for the terms of a manifold of the required sort. (They do not provide names for *all* of them; the number of names is $\aleph_0$, and therefore only a vanishing proportion of real numbers can be named—*i.e.* expressed by means of a formula of finite complexity which employs integers.) This possibility requires investigation.

The problem can be discussed equally well in two dimensions. In Gauss's theory of surfaces, a sphere and an ellipsoid, *e.g.*, are distinguishable by the fact that there is an irreducible difference between the formulæ for ds^2 which hold for the two surfaces when expressed in terms of two co-ordinates; this expresses the fact that the measure of curvature is constant in the case of the sphere, but not in the case of the ellipsoid. Yet from a purely ordinal point of view, such as that of *analysis situs*, the two figures are indistinguishable. What, exactly, is added to make the difference? This problem is essentially the same as that which arises in the general theory of relativity.

In part, the answer in this case is simple. What is added is the comparability of distances in different directions. So long as our apparatus is purely ordinal, we can say of three points which have the order ABC that B is nearer to A than C is, but we cannot say anything analogous of three points which are not in a row—I do not say "in a straight line," because the concept involved is more general, as will appear later. But although this is part of the answer, it does not seem to be the whole, since our relation S also enabled us to compare distances not having a common origin.

It seems that what distinguishes distance as required in geometry from such a relation as "subtending a given angle at a given point" is the absence of reference to anything external. When the distance between two points is equal to the distance between two others, we are supposed to have a fact which does not demand reference to some other point or points. In fact, this is the reason why the "interval" has been substituted for distance: the latter, as hitherto conceived, was found to depend upon the motion of the co-ordinate frame, and thus to be not an intrinsic geometrical relation. The distance, if it is to serve its purpose, must be a function of the two points exclusively, and must not involve any other geometrical data. Here, for relativity purposes, "geometry" includes "kinematics." The angle which two points subtend at a given point becomes a function of *three* points as soon as the given point is thought of as variable. There must be no such way of turning the distance between two points into a function involving other variables also.

I am not sure, however, whether it is necessary to introduce this somewhat difficult consideration. In ordinary geometry, the points at a given distance from a given point lie on the surface of a sphere; but if we define the distance PQ as the angle POQ, where O is a fixed point, the points at a given distance from P lie on a cone. Now a sphere and a cone are distinguishable in *analysis situs*. Thus the above undesirable definition could be excluded by insisting that points at a given distance from a given point are to form an oval figure. In relativity theory, this is not true of points having zero interval from a given point; indeed, it is only true when the interval concerned is space-like. But it is possible to specify the characteristics, for *analysis situs*, of the three-dimensional surface of constant distance from a given point. These might be added to the postulate that distance exists. Whether, in some such way, we could overcome the apparent necessity for distinguishing between a sphere and an ellipsoid, making the difference relative to the definition of distance, I do not feel sure, though obviously the question must be easily soluble.

Every principle of measurement which is to be used in practice must be such that important empirical laws are connected with measures. There will always be an infinite number of ways of correlating numbers with the members of a class whose cardinal number is less than or equal to $2^{\aleph_0}$. Some of these may be important, but most must be unimportant. Some conditions can be laid down. In the first place, the members of the class concerned may be obviously capable of an order which is causally important. If we take all the patches of colour that ever have been or will be perceived, they have in the first place an order in space-time, which is obviously important causally; in this order, no two of them occupy the same position—*i.e.* the relations concerned are all asymmetrical. But they have also an order as shades of colour and as of varying brightness. In this order there are symmetrical transitive relations—*e.g.* between two patches of exactly the same shade. Physics professes to correlate also these further characteristics of colours with spatio-temporal quantities such as wave-lengths. This would not be plausible if continuous alterations of quality were not correlated with continuous alterations in the correlated physical quantities. Whenever we notice a qualitative series, such as that of colours of the rainbow, we assume that it must have causal importance, and we insist that numbers used as measures shall have the same order as the qualities which they measure. The former is a postulate, the latter a convention. Both have proved highly successful, but neither is an *a priori* necessity.

There are orders which are obviously of no causal importance—*e.g.* alphabetical order among human beings. Human beings, like colours, have various orders that are causally important—the space-time order, order of height, weight, income, intelligence as measured by Professor X's tests, etc. But alphabetical order would never be thought important; no one would hope to found a biometric calculus upon a system in which a human being had co-ordinates depending upon the alphabetical order of his name. Generally speaking, it would seem that the simplest relations are the most important. Here I am using a purely logical test of simplicity: taking propositions in which the given relation occurs, there will be some having the smallest number of constituents compatible with the mention of that relation; and again, a relation may be a molecular compound of other relations—*i.e.* a disjunction, conjunction, negation, or complex of all these. A relation which is molecular has always a certain definite number of atoms; a relation which is not molecular is called atomic, and has then a definite number of terms in the simplest propositions in which it occurs. An atomic relation is simpler in proportion to the fewness of its terms; a molecular relation, in proportion to the fewness of its atoms. There is much empirical reason to think that the laws of a science become more important and comprehensive as the relations involved become simpler. The relation of a man to his name is of immense complexity, whereas we may suppose that the relation upon which interval depends is fairly simple. And the qualitative order of colours alluded to above is also simple, so long as we are thinking of colours as given in perception, not as interpreted in physics. Such simple relations should, as far as possible, be the basis for systems of measurement.

There is a traditional distinction between extensive and intensive quantities, which is somewhat misleading when taken seriously. The theory is that extensive quantities are composed of parts and intensive quantities are not. The only truly extensive quantities are numbers and classes. Where finite classes are concerned, the number of their terms may be taken as a measure of them, and they have parts corresponding to all smaller numbers. But in geometry we are never concerned with quantities which have parts. The number of points in a volume, whether large or small, is always $2^{\aleph_0}$ in the usual kinds of geometry; thus magnitude has nothing to do with number. Interval, as we have seen, is a relation, and smaller intervals are not parts of it. If AB and BC are equal intervals in a straight line, we say that the interval AC is double of each, and we think of it as the "sum" of AB and BC. But it is only by a convention, though an almost irresistible one, that we assign as the measure of AC a number double that which we assign as the measure of AB or of BC. And to say that AC is the "sum" of AB and BC is to say something very ambiguous, since the word "sum" has many meanings. When AB and BC are considered as vectors, we may say that AC is their sum even when they are not in one straight line. Again, given suitable definitions, we may say that the points between A and C are the sum (in the logical sense) of the points between A and B, and between B and C; this will only hold if ABC is a straight line. But the distance between A and C, considered as a relation, is not properly the "sum," in any recognized sense, of the distances AB, BC. Thus all geometrical quantities are "intensive." This shows that the distinction of intensive and extensive is unimportant.

In connection with interval, it is worth while to compare its formal characteristics with those of similarity. We saw that, in the generalized geometry with which Eddington ends, we want a relation of four neighbouring points, expressing the fact that they form a parallelogram. But we met with certain difficulties owing to the fact that this is only supposed to be possible for an infinitesimal quadrilateral, which is a figment of the mathematical imagination, and that it was not wholly easy to see how to substitute a procedure by means of limits. We were led to the suggestion that, instead of saying "*abcd* is a parallelogram," we should have to say "*abcd* is more nearly a parallelogram than *efgh.*" Perhaps this could be somewhat simplified. Suppose we say: "*abcd′* is more nearly a parallelogram than *abcd.*" And perhaps this could be still further simplified so as to take the form: "*cd′* is more like *ba* than *cd* is." We here suppose that between *any* two points there is a relation, which we will not call distance, but (say) "separation," and that this relation, like a shade of colour, is capable of a greater or less resemblance to another of the same kind. In a Euclidean space, two finite separations finitely separated may be exactly similar in the relevant respects; we then have a finite parallelogram. But in the generalized geometry that we are considering, we shall say that no two separations are *exactly* alike, though they are capable of indefinite approximation to exact likeness. Let us see how far this will take us.

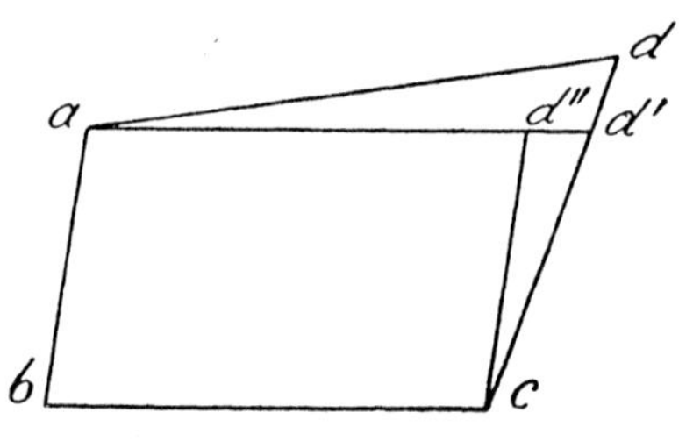

In the case of similarity, we have a relation which is capable of degrees, and may be called "quasi-transitive"—*i.e.* if A is very like B, and B is very like C, then A must be rather like C. This is just the sort of thing required for Weyl's geometry. Consider four points, a, b, c, d, and suppose that ab is rather like cd. Take a series of points forming a continuous route from c to d, without loops; this can be done by purely ordinal methods to be explained later. Suppose that among these points there are some, such as d', which make cd' more like ab than cd is. We may suppose that these points have a limit or last term, which we will call d'. We can then similarly proceed along ad' to a point d'' which gives cd'' more like ab than for any other point on ad'. We have then done nearly as well as possible, if not quite, with the three points a, b, c as starting-points. By means of suitable postulates, we could insure that a construction of the above sort, carried out repeatedly without changing the points a, b, c, should at last end with a definite point d_0 such that cd_0 is more like ab than any other distance from c is. We may call the figure $abcd_0$ a "quasi-parallelogram." Now let x_1, x_2, ... x_n, ... be a series of points on a route from b to a. Then proceed to take points y_1, y_2, ... between b and c on some route, and form the quasi-parallelograms having one corner at b, one corner at x_m and one at y_n, the fourth being called z_{mn}. If, as Weyl assumes, infinitesimal distances which have one end in common are comparable, this must be taken to mean that two small finite distances are capable of a resemblance which may be called "quasi-equality," which grows more nearly complete resemblance as the distance grows smaller. We may assume, as before, that, given a point x_1 and a definite route from b to c, there will be one definite point y_1 on this route such that by_1 is more nearly equal to bx_1 than is any other distance by on the route in question. We shall then say that bx_1 and by_1 are "quasi-equal." Take also x_1x_2, x_2x_3 ... quasi-equal, and y_1y_2, y_2y_3 ... quasi-equal. In this way we can construct a co-ordinate mesh with axes ba, bc. And we can now construct what will be in effect straight lines through b: take all the points z which are the corners opposite to b of quasi-parallelograms bx_mzy_n, for different initial points x_1, y_1, subject to quasi-equality between bx_1 and by_1. These points may be regarded as forming the quasi-straight line whose equation is $x/m=y/n$. (Irrationals can be dealt with by the usual methods.) This quasi-straight line will start from b in a

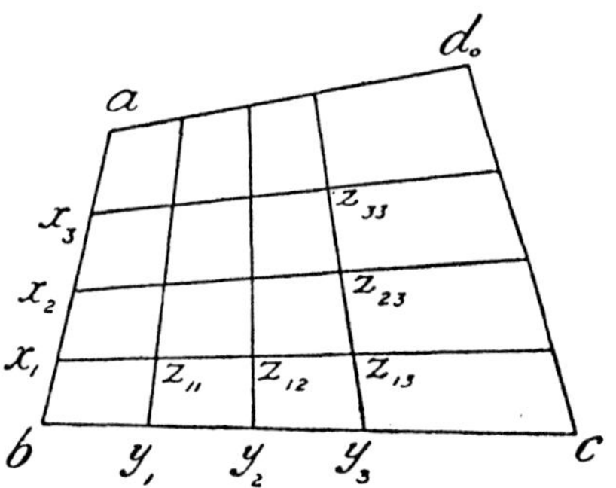

certain direction, and may, for differential purposes, be regarded as really a straight line. It is not worth while to proceed further, since it is obvious that we have the necessary material.

Degrees of similarity may be, in a sense, measured by quasi-transitiveness. Suppose that bx_1, y_1z_{11}, y_2z_{12}, . . . each have quasi-equality with the next. It may or may not happen that bx_1 has quasi-equality with y_nz_{1n}. One may presume that this will happen if bx_1 and by_1 are very small and n is not very large. Similarly, or rather *a fortiori*, we cannot infer that bx_m has quasi-equality with by_m. The larger the value of m for which such an inference remains true, the closer is the resemblance between bx_1 and by_1, or between bx_1 and y_1z_{11}. It is to be assumed that, by continually diminishing bx_1 and by_1, the number of steps for which the inference is permitted can be increased without finite limit.

If the above is in any degree valid, it would seem that, if space-time is continuous, spatio-temporal measurement depends theoretically upon qualitative similarity, capable of varying degrees, between relations of pairs of points. It is not suggested that the analysis cannot be carried further, but only that this is a valid stage in the process of explaining what is meant by the quantitative character of intervals and by their measurement as numerical multiples of units.

CHAPTER XIII

MATTER AND SPACE

COMMON sense starts with the notion that there is matter where we can get sensations of touch, but not elsewhere. Then it gets puzzled by wind, breath, clouds, etc., whence it is led to the conception of "spirit"—I speak etymologically. After "spirit" has been replaced by "gas," there is a further stage, that of the æther. Assuming the continuity of physical processes, there must be things happening between the earth and the sun when light travels from the sun to the earth; assuming the mediæval metaphysic of "substance," as all physicists did until recently, what is happening between the earth and the sun must be happening "in" or "to" a substance, which is called the æther.

Apart from metaphysical interpretations, what we may be said to know (using this word somewhat liberally) is that processes occur where there is no gross matter, and that these processes proceed, at least approximately, in accordance with Maxwell's equations. There does not seem any necessity to interpret these processes in terms of substance; indeed, I shall argue that processes associated with gross matter should also be interpreted so as not to involve substance. There must, however, remain a difference, expressible in physical terms, between regions where there is matter and other regions. In fact, we know the difference. The law of gravitation is different, and the laws of electromagnetism suffer a discontinuity when we reach the surface of an electron or proton. These differences, however, are not of a metaphysical kind. To the philosopher, the difference between "matter" and "empty space" is, I believe, merely a difference as to the causal laws governing successions of events, not a difference expressible as that between the presence or absence of substance, or as that between one kind of substance and another.

Physics, as such, should be satisfied when it has ascertained the equations according to which a process takes place, with just enough interpretation to know what experimental evidence confirms or confutes the equations. It is not necessary to the physicist to speculate as to the concrete character of the processes with which he deals, though hypotheses (false as well as true) on this subject may sometimes be a help to further valid generalizations. For the present, we are confining ourselves to the standpoint of physics. Whether anything further can be known or fruitfully conjectured is a matter which we shall discuss at a later stage. We want, therefore, to consider the difference in physical formulæ which is described as that between the presence and absence of matter, and also to consider briefly the difficulties as to the interchanges of energy between matter and empty space. I say "empty space" or "æther" indifferently; the difference seems to be merely one of words.

One way of approaching this subject is through the connection of mass with energy.* In elementary dynamics, the two are quite distinct, but nowadays they have become amalgamated. There are two kinds of mass involved in physics, of which one may be called the "invariant" mass, the other the "relative" mass. The latter is the mass obtained by measurement, when the body concerned may be moving relatively to the observer; the former is the mass obtained when the body is at rest relatively to the observer. If we call the invariant mass m and the relative mass M, then, taking the velocity of light as unity, if v is the velocity of the body relative to the observer, we have:

$$M = \frac{m}{\sqrt{1 - v^2}}.$$

* See Eddington, *op. cit.*, §§ 10, 11, 12.

Thus M increases as v increases; if v is the velocity of light, M becomes infinite if m is finite. In fact, the invariant mass of light is zero, and its relative mass is finite. Wherever energy is associated with matter, there is a finite invariant mass m; but where energy is in " empty space," m is zero. This might be regarded as a definition of the difference between matter and empty space.

It will be seen that, if v is small, so that v^4 and higher powers can be neglected, the above equation becomes approximately

$$M = m + \tfrac{1}{2}mv^2.$$

Now $\frac{1}{2}mv^2$ is the kinetic energy. Thus the change of M with changes of motion is the same as the change of the kinetic energy. But energy is fixed only to the extent of its changes, not in its absolute amount. Hence M may be identified with the energy. And this suggests further that the usual definition of energy is only an approximation, which holds when v is small. The accurate formula for energy is

$$\frac{m}{\sqrt{1 - v^2}}$$

—*i.e.* accurately the same as M.

The conservation of energy is the conservation of M, not of m; m also is approximately conserved, but not exactly. *E.g.* there is a loss of m when four protons and two electrons combine to form a helium nucleus. The term " invariant " refers to changes of co-ordinates, not to constancy throughout time.

It is necessary to say something about the difficulties of reconciling the laws governing the propagation of light with those governing interchanges of energy between light and atoms. On this subject the present position of physics is one of perplexity, aptly summarized by Dr Jeans in *Atomicity and Quanta* (Cambridge, 1926) and by Dr C. D. Ellis in *Nature*, June 26, 1926, pp. 895-7. The wave theory of light accounts adequately for all phenomena in which only light is concerned, such as interference and diffraction; but it fails to account for quantum phenomena such as the photo-electric effect (see Chapter IV.). On the other hand, theories which account for the quantum phenomena seem unable to account for the very things which the wave theory explains perfectly.

Some of the difficulties of the light-quantum theory are set forth as follows by Dr Jeans (*op. cit.*, pp. 29, 30):

"If, however, radiation is to be compared to rifle bullets, we know both the number and size of these bullets. We know, for instance, how much energy there is in a cubic centimetre of bright sunlight, and if this energy is the aggregate of the energies of individual quanta, we know the energy of each quantum (since we know the frequency of the light) and so can calculate the number of quanta in the cubic centimetre. The number is found to be about ten millions. By a similar calculation it is found that the light from a sixth magnitude star comprises only about one quantum per cubic metre, and the light from a sixteenth magnitude star, only about one quantum per ten thousand cubic metres. Thus if light travels in indivisible quanta like bullets, the quanta from a sixteenth magnitude star can only enter a terrestrial telescope at comparatively rare intervals, and it will be exceedingly rare for two or more quanta to be inside the telescope at the same time. A telescope of double the aperture ought to trap the quanta four times as frequently, but there should be no other difference. This, as Lorentz pointed out in 1906, is quite at variance with our everyday experience. When the light of a star passes through a telescope and impresses an image on a photographic plate, this image is not confined to a single molecule or to a close cluster of molecules as it would be if individual quanta left their marks like bullets on a target. An elaborate and extensive diffraction pattern is formed; the intensity of the pattern depends on the number of quanta, but its design depends on the diameter and also on the shape of the object-glass. Moreover, the design does not bear any resemblance whatever to the 'trial and error' design which is observed on a target battered by bullets. It seems impossible to reconcile this with the hypothesis that quanta travel like bullets directly from one atom of the star to one molecule of the photographic plate."

The difficulties of the wave-theory, on the other hand, are illustrated by Dr Ellis as follows:

" To take a definite case, suppose X-rays are incident on a plate of some material, then it is found that electrons are ejected from the plate with considerable velocities. The number of the electrons depends on the intensity of the X-rays and diminishes in the usual way as the plate is moved farther from the source of X-rays. The velocity or energy of each electron, however, does not vary, but depends only on the frequency of the X-rays. The electrons are found to have the same energy whether the material from which they come is close to the X-ray bulb or whether it is removed away to any distance.

" This is a result which is quite incompatible with the ordinary wave-theory of radiation, because as the distance from the source increases the radiation spreading out on all sides becomes weaker and weaker, the electric forces in the wave-front diminishing as the inverse square of the distance. The experimental result that the photo-electron always picks up the same amount of energy from the radiation could only be accounted for by giving it the power either to collect energy from a large volume or to collect energy for a long time. Both of these assumptions are unworkable, and the only conclusion is that the radiated energy must be localized in small bundles.

" This is the basis of the light-quantum theory. Light of frequency ν is considered to consist of small bundles or quanta of energy all identical and of magnitude $h\nu$, h being Planck's constant. These quanta travel through space, being unaffected by each other, and preserving their own individuality until they make a suitable collision with an atom."

After setting forth the difficulties encountered by this theory in regard to interference and diffraction, Dr Ellis proceeds to the very interesting suggestion made by Professor G. N. Lewis in *Nature*, February 13, 1926, p. 236. " It is a striking fact," says Dr Ellis, summarizing this suggestion, " that while all the theories are directed towards explaining the propagation of light, one theory suggesting that it occurs in the form of waves, the other in the form of corpuscles, yet light has never been observed in empty space. It is quite impossible to observe light in the course of propagation; the only events that can ever be detected are the emission and absorption of light. Until there is some atom to absorb the radiation we must be unaware of its existence. In other words, the difficulty of explaining the propagation of light may be because we are endeavouring to explain something about which we have no experimental evidence. It might be more correct to interpret the experimental facts quite directly and to say that one atom can transfer energy to another atom although they may be far apart, in a manner analogous to the transference of energy between two atoms which collide."

Professor Lewis's theory suggests that we should take seriously the fact that the interval between two parts of a light-ray is zero, so that its point of departure and its point of arrival may be regarded as, in some sense, in contact. In a passage quoted by Dr Ellis, he says:

"I shall make the contrary assumption that an atom never emits light except to another atom, and that in this process, which may rather be called a transmission than an emission, the atom which loses energy and the atom which gains energy play co-ordinate and symmetrical parts."

In a later letter to *Nature* (December 18, 1926), Professor Lewis suggests that light is carried by corpuscles of a new sort, which he calls "photons." He supposes that, when light radiates, what happens is that a photon travels; but at other times the photon is a structural element within an atom. The photon, he says, "is not light, but plays an essential part in every process of radiation." He assigns to the photon the following properties: "(1) In any isolated system the total number of photons is constant. (2) All radiant energy is carried by photons, the only difference between the radiation from a wireless station and from an X-ray tube being that the former emits a vastly greater number of photons, each carrying a very much smaller amount of energy. (3) All photons are intrinsically identical. . . . (4) The energy of an isolated photon, divided by the Planck constant, gives the frequency of the photon. . . . (5) All photons are alike in one property which has the dimensions of action or of angular momentum, and is invariant to a relativity transformation. (6) The condition that the frequency of a photon emitted by a certain system be equal to some physical frequency existing within that system, is not in general fulfilled, but comes nearer to fulfilment the lower the frequency is." Professor Lewis promises to deal with difficulties in the way of his hypothesis on a future occasion.

Professor Lewis's view is perhaps less radical than the view which it suggests—namely, that nothing whatever happens between the emission of light by one atom and its absorption by another. Whether this view is Professor Lewis's or not, it deserves to be considered, for although it is revolutionary, it may well prove to be right. If so, "empty space" is practically abolished. There will be need of a considerable labour if physics is to be re-written in accordance with this theory, but what is said about the necessary absence of evidence concerning light in transit is a powerful consideration. It is common in science to find hypotheses which, from a theoretical point of view, are unnecessarily complicated, because people cannot sufficiently divest themselves of common-sense prejudices. Why should we suppose that anything at all happens between the emission of light and its absorption? One might be inclined to attach weight to the fact that light travels with a certain velocity. But relativity has made this argument less convincing than it once was. Everything that has to do with the velocity of light is capable of being interpreted in a "Pickwickian" sense, and in any case our prejudices must be shocked. It is of course premature to adopt such an hypothesis definitively, and I shall continue to suppose that light does really travel across an intervening region. But it will be wise to remember the possibility, and to bear in mind the great changes in our imaginative picture of the world that are compatible with our existing physical knowledge.

The picture presented by this development of Professor Lewis's suggestion would be something like this: the world contains bits of matter (electrons and protons) possessing various amounts of energy. Sometimes energy is transferred from one of these bits of matter to another; usually this process has been thought to be casual, like the wandering of thistledown, but it is found to be more like the parcels post, in the sense that the energy has a definite destination. It is now suggested that there is no postman, because, if there were, he would be as magical as Santa Claus; the alternative is to suppose that the energy passes immediately from one piece of matter to another. It is true that, by the clock, there is a lapse of time between the departure of the energy from the source and its arrival at its destination. But there is no interval in the relativity sense, and the lapse of time will vary according to the co-ordinate system employed—*i.e.* according to the way in which the clock is moving. I do not know how the view we are considering will account for the time taken by a double journey to a reflector and back, which is not purely conventional. Nor do I know what will happen to the conservation of energy if light cannot be radiated into the void. This latter argument, however, is not serious, since light which never hits a piece of matter is in any case purely hypothetical. I am not sure, either, that the theory is intended to be as radical as I have suggested; perhaps it is only meant that light never starts on a journey without having a destination in view. In this form, however, the theory would seem scarcely credible: we should have to suppose that matter could exercise a mysterious attraction from a distance, which would undo the gain derived from Einstein's theory of gravitation. Perhaps the theory may have gained undue plausibility from a belief that the whole geometry of space-time depended upon interval, whereas in fact there is a space-time order which is not derivable from interval, and which, as presupposed in relativity theory, does not regard as contiguous parts of a light ray which would ordinarily be regarded as widely separated.* Perhaps it may be possible to avoid these difficulties, but, if so, a very great theoretical reconstruction will be necessary. Meanwhile it must be regarded as still possible that some less revolutionary theory may solve the difficulties connected with the interchange of energy between light and bodies.

There are three papers by Einstein which discuss the possibility of obtaining quantum laws as consequences of a modified relativity theory.† These papers do not arrive at any definite conclusion confidently asserted; but they suffice to show that the problem of combining quantum laws with those of gravitational and electromagnetic fields is not a hopeless one, a view which is strengthened by Mr L. V. King's theory alluded to above (Chapter IV.). So long as it is not known to be hopeless, it is perhaps rash to fly to heroic solutions of the problem. And it is as yet by no means universally admitted that the wave-theory of light is inadequate in its own domain; Dr Jeans (*loc. cit.*), for example, regards the hypothesis of light-quanta as unnecessary for reasons which demand serious consideration. We must therefore await further knowledge before venturing upon a definite opinion.

* On this matter, *cf.* Eddington, *op. cit.*, § 98 (pp. 224-6).

† *Bietet die Feldtheorie Möglichkeiten für die Lösung des Quantenproblems?* Sitzungsberichte der preussischen Akademie der Wissenschaften, 1923, pp. 359-64. *Quantentheorie des einatomigen idealen Gases.* *Ib.*, 1924, pp. 261-7, and 1925, pp. 3-14.

CHAPTER XIV

THE ABSTRACTNESS OF PHYSICS

BEFORE embarking upon the epistemological discussions which will concern us in Part II., it will be well to draw some morals from our previous chapters. Throughout these chapters, I have carefully abstained from speculations which would have taken us outside the domain of physics; in particular, I have not sought to interpret the mathematically fundamental notions of physics in terms of entities not directly amenable to ordinary mathematical treatment. It seemed desirable to be clear first as to what physics has to say, before undertaking either the epistemological criticism of the evidence or the metaphysical interpretation of the logically primitive apparatus of physics. This is the purpose of the present chapter.

Physics started historically, and still starts in the education of the young, with matters that seem thoroughly concrete. Levers and pulleys, falling bodies, collisions of billiard balls, etc., are all familiar in everyday life, and it is a pleasure to the scientifically-minded youth to find them amenable to mathematical treatment. But in proportion as physics increases the scope and power of its methods, in that same proportion it robs its subject-matter of concreteness. The extent to which this is the case is not always realized, at any rate in unprofessional moments, even by the physicist himself; he may tell you that he can " see " an electron hitting a screen, which is of course a telescoped expression for a complicated inference. Dr Whitehead has done more than any other author to show the need of undoing the abstractions of physics. For the moment, I am not concerned with this need, but with the abstractions themselves.

Let us take space, time, light and matter as illustrative of the gradually increasing abstractness of physics. These four notions are all extracted from common sense. We see objects spread out in space, we can feel their shapes with our fingers; we know what it is to walk to a neighbouring town or travel to a neighbouring country. All this makes "space" seem something familiar and easy, until, in the course of education, we learn the puzzles to which it has given rise. Time seems equally obvious: we remember past events in a time-order we notice day and night, summer and winter, youth and age, we know that history relates events of previous epochs, we insure our lives in the confident expectation that we shall die in the future. Light, again, seemed in no way mysterious to the author of Genesis, as, indeed, how should it to anyone who had experienced the difference between night and day? Matter was equally obvious: it was primarily anything that we could touch, though the first step towards mystification was taken when Empedocles included air. However, we are conscious of air in the form of wind and as something that fills our lungs, so that less effort was required to admit air among the elements than to exclude fire.

From this happy familiarity with the everyday world physics has been gradually driven by its own triumphs, like a monarch who has grown too grand to converse with his subjects. The space-time of relativity is very far removed from the space and time of our unscientific experience; yet even space-time is nearer to common sense than the conceptions towards which physics is tending. "Space and time," says Eddington,* "are only approximate conceptions, which must ultimately give way to a more general conception of the ordering of events in nature not expressible in terms of a fourfold co-ordinate system. It is in this direction that some physicists hope to find a solution of the contradictions of the quantum theory.

* *Op. cit.*, p. 225.

It is a fallacy to think that the conception of location in space-time based on the observation of large-scale phenomena can be applied unmodified to the happenings which involve only a small number of quanta. Assuming that this is the right solution it is useless to look for any means of introducing quantum phenomena into the later formulæ of our theory; these phenomena have been excluded at the outset by the adoption of a co-ordinate frame of reference." But even if space-time, as it appears in the general theory of relativity, were the last word as regards the physical order corresponding to our usual notions of space and time, it is evident that we should have travelled very far from those notions, and have arrived at a region in which pictorial imagination is useless.

The view of Locke, that the secondary qualities are subjective but not the primary qualities, was more or less compatible with physics until very recent times. There are spaces and times in our immediate experience, and there seemed no insuperable obstacle to identifying them with the spaces and times of the physical world. In regard to time, at least, practically no one doubted the rightness of this identification. There were doubts as regards space, but they came from psychologists rather than physicists. Now, however, both space and time, as they occur in immediate experience, are recognized by writers on relativity as something quite different from the space-time which physics requires. Locke's half-way house has therefore been definitely abandoned.

I come now to the relation of light as experienced to light in physics. Here the cleavage is older than in the case of space and time; indeed, it is already admitted in Locke's theory. It is impossible to exaggerate the importance of this cleavage in separating the world of physics from the world of common sense. With the exception of parts of our own body and bodies with which our own body is in contact, the objects which, according to common sense, we perceive, are known by means of light, sound, or odour. The last of these, though important to many species of animals, is relatively subordinate in the perceptions of human beings. Sound is less important than light, and in any case raises exactly the same problems in the present connection. We may therefore concentrate upon light as a source of our knowledge concerning the external world.

When we "see" an object, we *seem* to have immediate knowledge of something external to our own body. But physics says that a complicated process starts from the external object, travels across the intervening region, and at last reaches the eye. What goes on between the eye and the brain is a question for the physiologists, and what finally happens when we "see" is a question for the psychologist. But without troubling ourselves about what happens after the light reaches the eye, it is evident that what the physicist has to say is destructive of the common-sense notion of "seeing." It makes no difference, in this matter, which of the possible theories we adopt as to the physical character of light, since all equally make it something utterly different from what we see. The data of sight, analyzed as much as possible, resolve themselves into coloured shapes. But the physical analogue of a colour is a periodic process of a certain frequency relative to the eye of the observer. The physical world, it seems natural to infer, is destitute of colour. Moreover, the correspondence between colours and their physical counterparts is peculiar: colours are qualities, which are static while they last, whereas their counterparts are periodic processes, which are in the medium between the eye and the object which we say we "see." What happens in the object itself, if it shines by its own light, is the sort of thing considered in Bohr's theory: a sudden jump of an electron from one orbit to another. This is very unlike a sensation of (say) red. And what looks to the eye like a continuous red surface is supposed to be really a volume whose apparent colour is due to the fact that some of the electrons in it are jumping in a certain way. When we say they are "jumping," we are saying something too pictorial. What we mean is that they possess an unknown quality called "energy," which is a known function of a certain number of small integers, and that one or more of these integers have suddenly changed their values. It may be claimed as a merit in such theories as Professor Lewis's, considered in the preceding chapter, that it makes the connection between this process and the eye rather less indirect than it appears on the undulatory theory. But even then the sort of sudden transition contemplated by Bohr is very unlike the perception of a red patch: it is *prima facie* quite dissimilar in structure, and unknown as regards its intrinsic properties.

I come now to the most serious of our questions: How is matter to be understood in modern physics? Educated common sense regards matter as the cause of sensations; broadly speaking, sensations private to one person are caused by the matter of that person's body—*e.g.* headaches and toothaches—while sensations common to several, or of a sort which is common to several in suitable circumstances, are attributed to causes external to the bodies of the persons experiencing the sensations. (I am not at present attempting to make these statements exact, but merely to interpret what common sense would reply if questioned.) We recognize the "same" piece of matter on different occasions by similarity in its qualities, though we admit that this is a rough-and-ready test which may lead us astray. We think, however, that, if we had observed closely and continuously, we could have distinguished between two similar objects by means of continuity in their perceived spatial relations. The three-card trick illustrates what I mean: if we watch the performer carefully, we can tell which is the card we saw a moment ago, by means of the spatio-temporal continuity of its positions. What common sense assumes may be expressed, in language foreign to common sense, by saying: A piece of matter is manifested by sensible qualities whose variations are continuous, and whose sensible spatial relations to other such continuous series of qualities are continuous functions of the time. In practice, the changes of sensible quality are often so slow as to be negligible, and this greatly facilitates the task of common sense in recognizing the " same " object on two different occasions.

On the common-sense level, there are difficulties in certain cases: a drop, in a sensibly homogeneous fluid in which there is a current, cannot be distinguished at a later moment from another drop which was near it at the earlier moment. Combustion also offers difficulties to common sense. Both these matters can, however, be dealt with on a common-sense basis. A small solid object floating in the water will show which way the water is moving, and the smoke shows, more or less, what happens to an object which is burned. The elaboration immediately suggested leads on naturally to elementary physics and chemistry, where it is still assumed, at least tacitly, that the objects concerned are of the same sort as sensible objects, but rather smaller. Often they can actually be seen under the microscope. Imaginatively, we continue to attribute this continuity with sensible objects to our scientific objects, our electrons and protons, thus concealing from ourselves the highly abstract character of our assertions. At moments, we realize this abstractness; but it does not make its due impression, because imagination reasserts itself as soon as we are off our guard.

In theoretical physics, what is an electron, and how do we decide whether two events belong to the history of the same electron ? I am not asking how we decide in practice, but what is our theoretical definition. Ever since Minkowski, people have spoken of " world-lines," which are in fact the series of events constituting the history of one unit of matter.

but they have not always been as explicit as one could wish in telling us the criterion by which, in theory, it is decided that two events belong to one world-line. The test of identity between the parts of a world-line must obviously depend upon the laws of physics. These laws say that a material unit will move in such-and-such a way; inverting this statement, they say that what has moved in such-and-such a way is to count as one unit of matter. This is substantially the method pursued by Eddington. In Chapter IX. we considered the tensor

$$G_{\mu}^{\nu} - \tfrac{1}{2}g_{\mu}^{\nu}G,$$

which, as Eddington shows (§ 52), has the property of conservation—*i.e.* if the amount of it in any closed region varies, it does so by a flux across the boundaries. He identifies this quantity with matter, because of its property of conservation: "The quantity $G_{\mu}^{\nu} - \frac{1}{2}g_{\mu}^{\nu}G$ appearing in our theory is, on account of its property of conservation, now identified with matter, or rather with the mechanical abstraction of matter which comprises the measurable properties of mass, momentum and stress sufficing for all mechanical phenomena" (p. 146). And the above quantity, it will be remembered, is defined solely by means of the formula for small intervals. It will be admitted that matter, so defined, has become rather different from the matter in which common sense believes. If Dr Johnson had known Eddington's definition of matter, he might have been less satisfied with his practical refutation of Berkeley.

The exact form of Eddington's definition is not important for our present purposes; indeed, he himself somewhat generalizes it in a later passage. The point is that it is the *sort* of definition to which modern physics is bound to be led. Approximately, matter as conceived by common sense is conserved; wherever it appears to be destroyed or created, we can find ways of explaining away this appearance. Hence, as an ideal suggested by empirical facts, we adopt the view that matter is indestructible. We then turn round, and beginning from the formula for interval we construct a mathematical quantity which is indestructible. This, we say, we shall call "matter"; and no harm comes of our doing so. But whenever we take a step of this sort, we widen the gulf between mathematical physics and observation, and increase the problem of building a bridge between them. This problem has not been taken as seriously by physicists as it deserves to be taken. The reason is partly that it has arisen gradually. Physics and perception are like two people on opposite sides of a brook which slowly widens as they walk: at first it is easy to jump across, but imperceptibly it grows more difficult, and at last a vast labour is required to get from one side to the other. Another reason is that physiology and psychology, the two sciences concerned with perception, are less advanced than physics. The man accustomed to the beauty and exactitude of physics is liable to feel a kind of intellectual nausea when he finds himself among the uncertain and vague speculations of the less scientific sciences. He cannot be expected to admit that these sciences have a part to play in providing the premisses for his own precise mathematical deductions. Perhaps he is right, but *prima facie* physics, as an empirical study, derives its facts from perception, and cannot remain indifferent to any argument which throws doubt on the validity of perception, least of all when that argument is derived from physics itself. An argument designed to prove that a proposition is *false* is not invalidated by having that proposition among its premisses. Hence *if* modern physics invalidates perception as a source of knowledge about the external world, and yet depends upon perception, that is a valid argument against modern physics. I do not say that physics in fact has this defect, but I do say that a considerable labour of interpretation is necessary in order to show that it can be absolved in this respect. And it is because of the abstractness of physics, as developed by mathematicians, that this labour is required.

The inevitable specialism which is forced upon men of science by the very increase of scientific knowledge has had a good deal to do with obscuring this problem. Few men have been both physicists and physiologists. Helmholtz's researches concerning vision are a notable example of the combination of these studies, but there are not many others. Physiologists and psychologists are seldom well-informed in physics, and are apt to assume an old-fashioned physics which makes their problems look easier than they are. Moreover, even when the problem is realized, a man may not possess a mastery of the proper instrument for its solution—namely, mathematical logic. It is by means of mathematical logic that Dr Whitehead has been enabled to make his immense contribution to our problem. But, greatly as I admire his work, which I place far above anything else that has been written on the relation of abstract physics to the sensible world, I think there are points—and not unimportant points—where his methods break down for want of due attention to psychology and physiology. Moreover, there seem to be premisses in his construction which are derived rather from a metaphysic than from the actual needs of the problem. For these reasons, I venture to think that it is possible to obtain a solution less revolutionary than his, and somewhat simpler from a logical point of view. The solution, however, must wait until we have examined perception as a source of knowledge, which will be our topic in Part II. The metaphysic which reconciles the results of Part II. with the abstract physics which we have been considering in Part I. will be the subject of Part III.

PART II

PHYSICS AND PERCEPTION

CHAPTER XV

FROM PRIMITIVE PERCEPTION TO COMMON SENSE

In this Part, the subject with which we are concerned is the evidence for the truth of physics—not of this or that special result in physics, but of the general structure of the science. It is to be expected that the evidence will not be such as to give certainty, but at best such as to give probability; it is to be expected, also, that this probability may be increased by a suitable interpretation of physics, where "interpretation" is understood in the sense considered in Chapter I. We shall find it desirable to divide our problem into several parts, each of which will have an importance not confined to physics. There is need, first, to be clear as to what we mean by an empirical science, and what is the degree of certainty to be expected of it at the best. There is need to discuss what can be meant by "data," and to distinguish inferences, theories and hypotheses. We shall then discuss the causal theory of perception, and at the same time the philosophy called "phenomenalism." From these topics we shall pass to general discussion, first of cause, then of substance. This will lead us to the epistemological grounds for interpreting physics in accordance with neutral monism, and to the paramount importance of structure in scientific inference. We shall conclude with a definition of perception considered as affording the empirical data for physics, and with the consideration of phenomena analogous to perception in the non-mental world. But first of all it will be well to examine the historical development by means of which our problem has assumed its present form—both the prescientific development leading to common sense, and the scientific development leading from common sense to physics.

Common sense consists of a set of beliefs, or at least habits, which work well in practice except in situations which rarely occur. A savage may be puzzled by a box containing an unseen gyroscope, or by rails carrying an electric current; common sense has not prepared him for oddities of this sort. But a little familiarity enables a man to fit them into his common-sense world, and a mechanic soon learns their ways if he has occasion to do so. This illustrates the fact that there is no sharp line between science and common sense: both involve expectations, but those resulting from science are more accurate. It is possible to pursue science practically without any fundamental change from the metaphysic of common sense. But when theoretical science is taken seriously, it is found to involve a quite changed metaphysic, whose relation to that of common sense demands investigation. This will form the topic of the next chapter; in the present chapter, I shall consider the genesis of common sense, not in the race, since that is undiscoverable, but in the individual.

In studying infants, as in studying animals, we are compelled to confine ourselves to behaviouristic methods, whatever our views may be on the subject of behaviourism as a general principle in psychology. We can observe the bodily acts of young infants, but they cannot tell us their thoughts. At a low mental level, however, it is hardly profitable to distinguish between a belief and a habit of action. Beliefs, in the psychological sense, seem to emerge out of previously existing habits, and to be, at first, little more than verbal representations of habits formed before words could be uttered. There is therefore no great loss in being confined to behaviouristic methods when we are considering infants before the age at which they can speak.

It is of course obvious and generally recognized that very young infants do not possess the common-sense notion of an "object." This is by no means obvious with the young of some other kinds of animals—with chickens, for example. They possess, as instincts, useful ways of behaviour which in the human young are only learnt by experience; for example, they can pick up a grain which they see on the ground. The human infant has no such innate skill; for several months, it makes no attempt to touch what it sees. The "hand-eye co-ordination" comes as a result of experience. Some native aptitudes, of course, a new-born child does possess; for example, it can turn its eyes towards a bright light, though not very quickly or accurately. It has a reflex connected with sucking, but not a very intelligent one; indeed, it hardly amounts to more than the practice of trying to suck anything that comes in contact with the lips. Even in this respect, the human infant is inferior to the young of other mammals. We can say that certain stimuli rouse certain reflexes, but these are only just sufficient to keep the infant alive with the help of maternal care.

In this primitive condition, the infant obviously has no conception of an "object." An "object," for common sense, is something having a certain degree of permanence, and connected with several kinds of sensation. This involves something like memory, to give rise to the idea of permanence, or rather, at first, to the feeling of recognition; and it involves experience, to give to one sensory stimulus a reaction originally associated with another. In infants, the most important factor in forming the common-sense notion of an object is the hand-eye co-ordination, the discovery that it is possible, often, to grasp what is seen. In this way, visual and tactual spaces become correlated, which is one of the most important steps in the mental growth of an infant.

At this point, it is important to be clear as to the difference between "space" in psychology and "space" in physics. There is undoubtedly a connection between the two, which it will be part of our business to make clear at a later stage. But the connection is very round-about and inferential. At the outset, it is much more useful to realize the difference between them than the connection, since much confusion of thought arises from supposing the connection to be closer than it is. In physics there is only one space, while in psychology there are several for each individual; these can, it is true, be reduced by manipulation to one for each individual, but they cannot be reduced further without introducing obscurities that it is impossible to dissipate. The space containing my visual objects has no point in common with the space containing yours, since no visual object in my world is precisely identical with one in yours. And the amalgamation of the spaces of my different senses into one space is a piece of early science, performed by the infant at about the age of three months. Dr Whitehead, who is anxious to bridge the gulf between perception and physics, seems to me to make his task too easy where space is concerned. For example, he says:*

"The current doctrine of different kinds of space—tactual space, visual space, and so on—arises entirely from the error of deducing space from the relations between figures. With such a procedure, since there are different types of figures for different types of sense, evidently there must be different types of space for different types of sense. And the demand created the supply. If, however, the modern assimilation of space and time is to hold, we must go further and admit different kinds of time for different kinds of sense—namely, a tactual time, a visual time, and so on. If this be allowed, it is difficult to understand how the disjecta membra of our perceptual experience manage to collect themselves into a common world. For example, it would require a pre-established harmony to secure that the visual newspaper was delivered at the visual time of the visual breakfast in the visual room, and also the tactual newspaper was delivered at the tactual time of the tactual breakfast in the tactual room. It is difficult enough for the plain man—such as the present author—to accept the miracle of getting the two

* *The Principles of Natural Knowledge*, pp. 193-4.

newspapers into the two rooms daily with such admirable exactitude at the same time. But the additional miracle introduced by the two times is really incredible."

This passage is so pleasant that I hate to criticize it. But I do not know how else to make clear where I differ from Dr Whitehead. There is first a purely verbal question to be cleared up. Dr Whitehead says it is an error to deduce space from the relations between figures. It is certainly an error to deduce *physical* space in this way, but with psychological space the matter is different. There certainly are perceived relations between figures, and these perceived relations are part of our perceptual data in physics. Whether they are to be said to constitute a space or not, is a verbal question. Psychologists, as a rule, find it convenient to say so; but the matter is unimportant. When this question has been cleared away, however, there remain others which are vital to an understanding of the relation between physics and perception.

Take, first, the question of the two times. As will appear when we come to the causal theory of perception, the whole of my perceptual world is, from the standpoint of physics, in my head; any two events which I experience together overlap in physical space, and all of them together, in *physical* space, occupy a volume smaller than my head, since it certainly does not include the hair, skull, teeth, etc. Consequently, on relativity principles, there is no question of two times, since this only arises for events which are spatially separated in physical space.

As for the necessity of distinguishing tactual and visual space: there are perceived relations between objects seen simultaneously, and also between objects touched simultaneously, and these relations are part of the crude material out of which we construct our notion of space. *These* relations cannot hold between a visual and a tactual percept. But there are other relations which do hold—namely, those of correlation: when I see my hand in contact with a visual object I feel it in contact with a tactual object, and moreover the visual and the tactual object have certain relations to each other—*e.g.* where we see a corner we get a tactual sensation of sharpness. All this, however, is learnt by experience; that is to say, we learn the *laws* of the correlation by experience. The infant can be seen learning them. One may call these laws " pre-established harmonies," but they are no more so than any other scientific laws. Unless we are going to say that all laws of nature must be demonstrable by pure logic, which is hardly conceivable nowadays, we must admit that there are co-existences and sequences which we expect on a basis of past experience, in spite of the fact that their failure would not be logically impossible. And the correlation of visual and tactual sensations is a case of this sort.

It is sometimes suggested, in such cases, that the correlated occurrences are merely different manifestations of one and the same entity. This is, in fact, the view of common sense, which holds that it can both see and touch the same object. I have no objection whatever to this way of speaking, and I do not deny that, rightly interpreted, it may express a correct view. But it remains nevertheless true that the entity said to be manifested is inferred from experience of a correlation, and that the percepts correlated are not *logically* interconnected, but only empirically. We have *v*, a visual percept, and at the same time *t*, a tactual percept. Each rouses appropriate reflexes, and, owing to their frequently occurring together, it happens in time that each rouses also the reflexes appropriate to the other. This practical induction occurs before the child has reflected that the two are correlated; indeed, unless he becomes a learned man he probably never realizes the correlation of *v* and *t*. But as soon as we reflect upon the matter we can see that there is no *necessary* correlation. It fails with blind men, and with men whose fingers have been anæsthetized.

In general, however, the correlation holds good. Common sense explains it by regarding both touch and sight as ways of getting to know an object which is at once tangible and visible. In the language of the causal theory of perception, we say that *v* and *t* have a common cause, in general external to the body. I do not wish to deny this, but only to point out that, when we are considering the grounds of our knowledge, we cannot say that we know of the correlation because we know of the common external cause. The order in knowledge is the opposite: we have evidence for the correlation in our experience, and we infer* the common cause from the correlation, so that the common cause cannot have more certainty than the correlation, which is its premiss. From a behaviouristic point of view, the infant " knows " the correlation when either stimulus calls out the response originally appropriate to the other.

We must here guard against a small possible misunderstanding. If *v* and *t* are invariably correlated, it may be said, it is impossible that one should occur without the other, and therefore there can be no means of judging whether one alone would elicit the response belonging to the other. In fact, the matter is not quite so simple as we have been taking it to be. What we learn by infantile experience is not that *v* and *t* are *always* correlated; it is possible to touch in the dark, or with the eyes shut, and it is possible to see without touching. What we learn is that the correlation can be brought about easily in many cases. Movements of the eye will usually give a visual sensation corresponding to a previously uncorrelated tactual sensation, and movements of the hand (or other part of the body) will, in a certain proportion of cases, give a tactual sensation corresponding to a previously uncorrelated visual sensation. Children practising the hand-eye co-ordination attempt to grasp objects not within their reach; it is only

* I am here using the word " infer " in a behaviouristic sense.

gradually that distance comes to be judged more or less correctly. When objects are not within our grasp, a new correlation comes into play—namely, between the visual sensation and the journey required to bring the object within our reach. Unfamiliar circumstances will cause even adults to make mistakes—for example, that of underestimating the depth of objects under water. Great distances remain permanently beyond the scope of common sense: only science can assure us that the sun is farther off than the moon.

What we can observe the infant learning is the bodily acts which will, in fact, reinforce a percept of one sense by a percept of another; more particularly he learns to touch what he sees—*i.e.* to procure for himself a correlated pair v, t, instead of the isolated v. Similarly he learns to look round when he hears a voice, and so on. All this implies that he has, so far as action is concerned, the notion of a physical object, as something capable of affecting several senses simultaneously. The element of recognition is logically separable, and arises somewhat earlier.

These motor habits are essential in generating common-sense beliefs, which arise at a much later stage of mental growth. Common sense, in its more primitive form, is hardly aware that there is such an occurrence as perceiving; it is only aware of the perceived object. And by the time that even the most rudimentary reflection begins, each sense calls out responses connected with other senses, so that even when, from the standpoint of external stimulus, only one sense is affected, the experience has the massiveness of something in which several senses are involved. See, for example, the pictures in Köhler's *Mentality of Apes*: here we see chimpanzees which are watching others with sympathetic movements of the arms that indicate stimulation of bodily feelings connected with balance, although the sole stimulus is visual. This accounts for the fact that common sense can so confidently identify an object touched and not seen with an object seen but not touched—*e.g.* the cricket-ball now successfully caught and the same ball as it flew through the air. The reason is that the experience is always richer than the sensory stimulus alone would warrant: it contains always responses arising from physiological experience of past correlations. If an adult were to hear a donkey's bray for the first time, without having previously known that there was an animal which made that noise, his experience would be amazingly unlike that of a normal adult in the same circumstances.

Common sense does not initially distinguish as sharply as civilized nations do between persons, animals, and things. Primitive religion affords abundant evidence of this. A thing, like an animal, has a sort of power residing within it: it may fall on your head, roll over in the wind, and so on. It is only gradually that inanimate objects become sharply separated from people, through the observation that their actions have no purpose. But animals are not separable from people on this ground, and are in fact thought by savages to be much more intelligent than they are.

Commón sense is, in most respects, naively realistic: it believes that, as a rule, our perceptions show us objects as they really are. It is able to hold this view because of the mass of experience which, in each individual, precedes the common-sense outlook. We do not think a distant person smaller than a person near at hand; we do not judge circular objects seen sideways to be elliptic; and so on. All this is, for common sense, part of the perception; it may be doubted whether it is not so also for psychology. But it is certainly not part of the infant's initial perceptive apparatus: it is something which the infant has to learn. Some of it is learnt after the beginnings of speech have been acquired—particularly a right judgment as to the size of distant objects. But at any rate by the time a child is three years old he has acquired the common-sense outlook. That is to say, his immediate reaction to a sensory stimulus involves a great deal of previous experience, and is such as to enable him to arrive, without any mental process, at a far more objective view of what he perceives than was possible at birth. I mean here by "objective" not anything metaphysical, but merely "agreeing with the testimony of others." It would be a complete mistake to suppose that, in an adult, there is first an experience corresponding to the bare sensory stimulus, and then an inference to that of which it is a sign. This may occur in certain cases, for example, if we watch a man drawing a face in an apparently haphazard manner, and do not realize till the last moment that a face is being intended. But such an experience is quite unlike normal perception, where the "inference," in the only sense in which it can be said to exist, is physiological, or at any rate not discoverable by introspection. It is because the sensory stimulus is able to lead us, without any mental intermediary, to an object practically identical with that perceived by others in our neighbourhood, that we are able to adopt the common-sense belief that we actually perceive external objects.

The notion of cause is part of the apparatus of common sense. I do not think it would be true to say that common sense regards objects as the causes of our perceptions; it would not, unless challenged, think of bringing in causation in this connection. It looks for causes when it is surprised, not when an occurrence seems perfectly natural. It demands causes for a mirage, a reflexion, a dream, an earthquake, a plague, and so on, but not for the ordinary course of nature. And the cause which it looks for, wherever the event concerned has great emotional interest, is pretty sure to be animistic: the anger of the gods, or something analogous. The idea of universal causation, and of causation divorced from purpose, belongs to a later stage of mental development, and marks the beginnings of philosophy and science.

Substance is a category which comes naturally to common sense, though without the attribute of indestructibility added by the metaphysicians—but as to this perhaps diverse opinions are possible. One would be inclined to suppose that common sense regards fire as destroying what it burns; but the Chinese, when they had made a solemn covenant, used to burn it, in order that the gods might take cognizance of it through the smoke. (A copy was kept for terrestrial purposes.) And races that practise cremation do not, as a rule, suppose that they are totally destroying the body. On the other hand, there has existed a religious prejudice against cremation which implied the belief that the body was thereby totally annihilated. I think one must conclude, therefore, that the attitude of common sense as to the indestructibility of substance is vacillating; on the whole, the success of physics in providing immortal material units represents a triumph of the philosopher over the plain man.

Substance, whether indestructible or not, is of great importance in primitive thought, and dominates syntax, through which it has dominated philosophy down to our own day. At a primitive stage, there is no distinction between " substance " and " thing "; both express, first in language and then in thought, the emotion of recognition. To an infant, recognition is a very strong emotion, particularly when connected with something agreeable or disagreeable. When the infant begins to use words, it applies the same word to percepts on two occasions, if the second rouses the emotion of recognition associated with memory of the first, or perhaps merely with the word which was learnt in presence of the first. (When I say that the infant uses the " same " word, I mean that he makes closely similar noises.) Using a given word as a response to stimuli of a certain kind is a motor habit, like reaching for the bottle. Two percepts to which the same word applies are thought to be identical, unless both can be present at once; this characteristic distinguishes general names from proper names. The basis of this whole process is the emotion of recognition. When the process, as a learning of motor habits, is complete, and reflection upon it begins, identity of name is taken to indicate identity of substance—in one sense in the case of proper names, in another sense in the case of names applicable to two or more simultaneous percepts—*i.e.* general names (Platonic ideas, universals). Throughout, language comes first and thought follows in its footsteps. And language is governed largely by physiological causation.

A substance or thing is supposed to be identical at different times, although its properties may change. John Jones is the same person throughout his life, although he grows from childhood to manhood, is sometimes pleased and sometimes cross, sometimes awake and sometimes asleep. Primarily, he is considered to be the same person because he has the same name. But the name, like the person, is not exactly the same on different occasions; it may be spoken loud or soft, quickly or slowly. These differences, however, are too slight to prevent recognition, except on rare occasions—*e.g.* when the name is pronounced very badly by a foreigner; one of the merits of names is that they change less than the person named.

The conception of substantial identity with varying properties is embedded in language, in common sense, and in metaphysics. To my mind, it is useful in practice, but harmful in theory. It is harmful, I mean, if taken as metaphysically ultimate: what appears as one substance with changing states should, I maintain, be conceived as a series of occurrences linked together in some important way. I will not yet argue this view. It would have been utterly foreign to physics until the substitution of space-time for space and time, with the corresponding substitution of a four-dimensional continuum of events for the older conception of persistent material units moving in a three-dimensional space. But the older conception still appears the natural one to apply to electrons and protons, so that physics may be said to have, at the moment, two different points of view on this issue. For the present, I am not concerned to criticize the notion of substance, but only to show its genesis, which I take to be derived from the pre-human emotion which we reflectively call "recognition," though it has not, originally, the definite cognitive character attached to the word when applied to the mental processes of an adult human being.

Induction, like substance, plays a large part in common sense, and has a basis which is primarily physiological. I am not at present discussing the validity of induction, but the cause of the practice of induction among animals, children, and savages. Of course the validity of induction is really assumed in such a discussion, since, without it, causes cannot be discovered. But we do not assume the validity of the primitive inductions which we are discussing; we assume only that there is *some* valid form of induction. Throughout genetic psychology we assume the validity of ordinary scientific procedure. If this assumption were to lead us to views on genetic psychology which threw doubt on the validity of scientific procedure, that would constitute a *reductio ad absurdum*, which would destroy genetic psychology along with the rest. Therefore, whenever some obviously invalid process is said to be the psychological source of a method essential to science, we must suppose, unless we are to embrace complete scepticism, that there is some valid process which, in most of the cases to which the invalid process is applied by unscientific people, gives rather similar results. All this has perhaps only a pragmatic justification, but whether this is the case cannot be decided *ab initio*. The real utility of investigating crude primitive forms of inference is that the contrast between them and current scientific inference may suggest directions in which the latter is capable of still further improvement. The direct logical importance of investigations into the origins of our mental processes is *nil*, but the importance as a means of stimulating imagination in the formation of hypotheses may be considerable. It is for this reason that the topics of the present chapter form a useful introduction to those which form our proper subject-matter.

The source of induction, speaking historically, is the general law of what Dr J. B. Watson calls " learned reactions." In its schematic simplicity, this law is as follows: If a stimulus S to a living body of an animal produces a reaction R, and a stimulus S' produces a reaction R', then if S and S' are applied together, there is a tendency for S alone, afterwards, to produce R' as well as R. *E.g.* if you expose a person frequently to a certain loud noise and a bright light simultaneously, after a while the loud noise alone will cause his pupils to contract. It is obvious that the *practice* of induction is simply the application of this law to cognitive reactions. If you have frequently heard the words " there's Jones " when you could see Jones, these words will in the end cause you to believe that Jones is present even if, for the moment, you do not see him. This form of induction is involved in understanding speech. It is obvious that, in its cruder forms, induction may give rise to false beliefs as well as to true ones; scientific methodology has to seek a form of induction which shall make false inferences much rarer than true ones. If such a form can be found, a man may train himself, in his professional activity, to abstain from the more primitive forms. But as an ordinary mortal he could not survive for a day if he refused to trust to what we may call physiological induction, which stores up in the body the lessons of past experience. In practice, a nearly instantaneous method of inference which is right nine times out of ten is preferable to a slow method which is always right. A man who subjected all his food to chemical analysis before eating it would avoid being poisoned, but would also fail to be adequately nourished.

Throughout the development of theory, great intellectual changes have been repeatedly necessitated by errors which were very small from the standpoint of practice. The theory of relativity is a remarkable instance of this: an immense reconstruction has been made to meet discrepancies which could only be detected by the most delicate measurements. The further science advances, the more minute become the facts which it cannot yet assimilate. Common sense does well enough for most of the needs of a pre-industrial community, but not for the construction of a dynamo or a wireless station. For these, we have to advance to the standpoint of pre-relativity physics. Machines involving relativity physics do not yet exist, but presumably they will some day. This, however, is beside the point. The point is, that a small discrepancy between theory and observation may indicate a large error in theory. Take, *e.g.*, naive realism and the velocity of light, the latter from a pre-relativity point of view. The supposition of common sense and naive realism, that we see the actual physical object, is very hard to reconcile with the scientific view that our perception occurs somewhat later than the emission of light by the object; and this difficulty is not overcome by the fact that the time involved, like the notorious baby, is a very little one. We cannot therefore argue from the practical success of common sense to its approximate theoretical accuracy, but only to a certain rough correspondence between its commoner inferences and those permitted by a correct theory. If physics has had to desert common sense, that is no reason for finding fault with physics.

CHAPTER XVI

FROM COMMON SENSE TO PHYSICS

It was in the seventeenth century that the scientific outlook, as opposed to that of common sense, first became important. It had existed in individuals among the Greeks, but it had not been able to point to sufficiently great achievements to impress the general educated public. It was in the seventeenth century that science began to win spectacular victories, and to develop an outlook definitely different, in certain important respects, from that of common sense. The historical aspects of this change have been set forth by Dr Whitehead in his *Science and the Modern World*, particularly in the chapter on "The Century of Genius," so admirably that it would be foolish to attempt to cover the ground again. I shall therefore select only certain topics which are important in relation to subsequent chapters.

The chief thing that happened in the seventeenth century, from our point of view, was the divorce between perception and matter, which occupied all the philosophers from Descartes to Berkeley, leading the latter to deny matter, while it had, in effect, led Leibniz to deny perception.

Common sense believes that there is interaction between mind and matter: when a stone hits us our mind feels pain, and when we will to throw a stone it moves. The development of physics made matter seem causally self-contained: it appeared that there were always physical causes for the movements of matter, so that volitions must be otiose. Descartes, believing in the conservation of *vis viva*, but ignorant of the conservation of momentum, thought that the mind could influence the direction of the motion of the animal spirits, but not its amount. This half-way house had to be abandoned by his followers, owing to the discovery of the conservation of momentum. They therefore decided that mind can never influence matter. They also decided that matter can never influence mind. This latter view was not based directly upon science, but upon the metaphysic which had been invented to explain away the apparent influences of mind on matter. To suppose that the movement of my arm is not caused by my volition is to suppose something very odd; it is no odder to suppose that the perception of my arm is not caused by my arm. The view that there were two substances, mind and matter, and that neither could act upon the other, explained the causal independence of the physical world, and entailed that of the mental world. Thus mind and matter became very widely separated—much more so than they had been before the rise of modern physics.

All modern philosophy before Kant is dominated by this problem, for which a variety of solutions were offered. Spinoza held that there was only one substance, whose only *known* attributes were thought and extension, which ran parallel without interaction, like the two perfect clocks of the occasionalists. Leibniz believed in an immense number of substances, all causally independent of each other, but all running parallel in virtue of a pre-established harmony; these substances were all minds, more or less developed, and matter was only a confused way of " perceiving " a number of substances. The word " perceiving " has, in Leibniz's philosophy, a peculiar meaning, derived from parallelism and from the notion of " mirroring the universe." Without attempting to adhere closely to Leibniz's own words, we may set forth the view which is implied in his system, whether he held it in its entirety or not, as follows: Each monad, at each moment, is in an infinitely complex state, which is capable of a one-one correspondence with the state of each other monad at that moment. (This is the pre-established harmony.) The differences between the states of different monads are like the differences between the aspects of a given object from different places, and are compared by Leibniz to differences of perspective or point of view. These differences are capable of arrangement in a three-dimensional order, so that the monads form a pattern which changes with the time. In addition to the one-one correspondences between the monads, there is a one-one correspondence between the state of each monad and the pattern formed by all the monads (mirroring the world). It will be seen that the latter logically implies the former: if each monad always mirrors the world, each is always in harmony with every other. Let us take a mathematical analogy: suppose the states of the mth monad at a given moment are represented by the numbers:

$$m - 1, m - 1/2, m - 1/3, \dots$$

then there is a one-one correspondence between these states and those of the nth monad, which are:

$$n - 1, n - 1/2, n - 1/3, \dots$$

and there is also a one-one correspondence between the states of each monad and the series:

$$1, 2, 3, \dots m, \dots n, \dots$$

which may be taken to be the series of monads. Substitute three continuous co-ordinates for one discrete co-ordinate, and we get a mathematical representation of Leibniz's world.

The obvious difficulty in this system was that no conceivable reason could be given for supposing that a monad mirrored the world. Leibniz himself was one monad, and, on his own theory, would have had exactly the same life if he had been the only monad, since the monads were "windowless." He could not therefore give any grounds against solipsism except some rather far-fetched arguments derived from theology and God's "metaphysical perfection." This defect was due to his theory of causality, which was an outcome of the Cartesian denial that one substance could act upon another, which in turn was inspired by the success of physics in establishing purely physical causal laws which seemed to account for all the motions of matter. In spite of this glaring defect, I have lingered on Leibniz's system, because I believe that it contains hints for a metaphysic compatible with modern physics and with psychology, although of course it will require very serious modifications.

The problem of perception remained unsolved, although it was one of the main pre-occupations of philosophers. Locke, important as he was, did not contribute much on this question, except his theory that primary qualities are objective and secondary qualities subjective; but his *Essay* led others to theories which have remained important. Berkeley discarded the material world, though he need not have discarded physics, since the formulæ of physics may perfectly well be applicable to collections of mental events, as Leibniz supposed. Berkeley does not seem to have been influenced by the argument which affected the Cartesians—namely, the supposed impossibility of interaction between mind and matter. What influenced Berkeley was rather the epistemological argument, that everything with which we are acquainted is a mental event, and there is no valid reason for inferring that there are events of quite another kind. This type of argument is, I think, new in Berkeley, when regarded as a source of metaphysics; in another form, it achieved fame through Kant. Hume carried the same type of reasoning much further than Berkeley did, since he was content to remain sceptical, whereas Berkeley employed scepticism about matter as a support of religion, and therefore had to limit the scope of his criticism of what passed as knowledge. Hume's criticism of the notion of cause cut at the root of science, and demanded an answer imperatively. Of course innumerable answers were forthcoming, but I cannot persuade myself that any of them were in any degree valid, not even that of Kant. I do not wish, however, to discuss at this moment any philosophy which has still a more than historical interest, as is the case with Berkeley, Hume, and Kant. Let us therefore return from this excursus to topics more intimately connected with science.

The profound and lasting effect of Cartesianism upon the outlook of philosophers and men of science was to widen the gulf between mind and matter. Physicists were satisfied with the view that their science could be pursued independently of considerations concerned with mind, and contentedly left the philosophers to wrangle, under the impression that philosophy did not matter to them. For a time, from the point of view of the progress of science, there was much truth in this view; but in the long run science cannot shut its eyes to problems which are logically relevant to its investigations. It may be admitted that most of what has passed for philosophy would not have been very useful to the men of science; but that was chiefly because philosophy was no longer being created by men like Descartes and Leibniz, who were of supreme eminence in science as well. It may be hoped that this state of affairs is coming to an end.

The "matter" of the Cartesians, owing to their denial of interaction between mind and matter, should have been just as abstract, and just as purely mathematical, as in the most modern physics. But in fact this was not the case: the technique of the period still depended upon notions which had an immediate basis in our own experience. We may perhaps distinguish three sorts of physics, in relation to the sense-experiences from which their ideas are derived: I will call them muscular physics, touch physics, and sight physics respectively. Of course no one of them has ever existed in isolation: actual physics has always been a mixture of the three.

But it will be a help in analysis to imagine a separation of each from the others, and ask ourselves which elements in actual physics belong to the first, which to the second, and which to the third. Broadly we may say that sight-physics has more and more predominated, and has achieved an almost complete victory over the others in the theory of relativity.

Muscular physics is embodied in the idea of "force." Newton evidently thought of force as a *vera causa*, not as a mere term in a mathematical equation. This was natural; we all know the experience of "exerting force," and are aware that it is connected with setting bodies in motion. By a sort of unconscious animism, physicists supposed that something analogous occurs whenever one body sets another in motion. Unfortunately for dynamics we have the experience of "exerting force" when we merely cause a body to preserve a constant velocity, as in dragging a weight along a road; this misled Aristotle into thinking that force was to be regarded as the cause of velocity, not of acceleration, a mistake first corrected by Galileo—though Leonardo came very near seeing the truth. It may be said: if force is a mathematical fiction, how can it be more "true" to regard it as proportional to the acceleration than to regard it as proportional to the velocity? The reason is that laws can be found connecting force with the situation of a body relative to other bodies, if force is defined as Galileo defined it, but not if it is defined as Aristotle defined it. Galileo's discovery that falling bodies have a constant acceleration, which is the same for all (*in vacuo*), is a very simple instance. More generally we may say: The laws of physics are, as a rule, differential equations of the second order—with respect to time in Newtonian physics, and with respect to interval in the physics of Einstein. This is a very different notion from that of force as derived from experience of muscular exertion; yet the one has led to the other by an evolution containing many intermediate links.

Touch-physics has led to the passion for conceiving the world as composed of billiard balls—a passion which existed already in the Greek atomists. We know what it is to bump into people, or to have them bump into us; we know that when this happens motion is communicated without the exercise of volition. Billiard balls exhibit the phenomena concerned in the best form for elementary mathematical manipulation. The way billiard balls move when they hit each other is not at all surprising; on the contrary, in a general way it is such as everyone would expect. If all the world consisted of billiard balls, it would be what is called "intelligible"—*i.e.* it would never surprise us sufficiently to make us realize that we do not understand it. The conservation of momentum, which is exemplified in the impacts of billiard balls, seemed to give an admirably simple view of the whole occurrence. We can regard momentum as "quantity of motion," and say that in an impact a certain quantity of motion is interchanged between two bodies, just as nowadays electrons are exchanged when one body becomes positively electrified and another negatively. This view was preferable to that which used force, because it did not seem to demand of matter anything even remotely analogous to volition; it was therefore beloved of pre-Newtonian materialism. It has, however, completely disappeared from modern notions of the structure of matter. The "atoms" which are believed to exist—electrons and protons—never come into contact, but move as if they exerted attractions and repulsions at a distance; these, however, are explained as due to something transmitted through the intervening medium. What has remained from touch-physics is an objection to "action at a distance." But this objection can hardly be now attributed to an *a priori* prejudice; it is rather the outcome of experiment. We believe that, when one body seems to influence another at a distance, this is either capable of being explained away, or is attributed to the continuous passage of energy across the space between the two bodies; but we believe this because it is the view which fits best with known facts, not because it seems the only "intelligible" view. The latter opinion is no doubt widely held, but is not required to justify existing physical theories.

Sight-physics has inevitably been dominant in astronomy, owing to the fact that sight is the only sense by means of which we have cognizance of the heavenly bodies. So long as we only see a motion, we are not conscious of anything analogous to force. The fact that gravitation remained so long unexplained may have stimulated the desire of theoretical physicists to develop their subject without the notion of "force," since the "force" of gravitation remained totally obscure. Sight-physics also had the advantage that it dealt with a wider range of phenomena than were included in dynamics, since it included everything to do with light. Thus physics came more and more to use only such notions as were intelligible in terms of visual data. Mass, it is true, remained from another order of ideas. Obviously the sensational source of the idea of mass is the feeling of weight. But even mass has gradually yielded. On the one hand, it is less fundamental than it formerly seemed; on the other hand, it can be inferred from optical data, by the deflection from a straight line which a body suffers in a known field of force. (Consider methods of determining the apparent masses of α and β particles.) Sight-physics also makes the relativity of motion much more evident than either of the other kinds. A train exerts force, and a railway station does not, so that, from this point of view, it seems natural and right to say that the train is "really" moving while the station is "really" at rest. But from a visual point of view the appearance of the station from the train is exactly correlative to that of the train from the station.

In the visual world, quite independently of the velocity of light, a rapid movement can be produced by a very small "force"—for instance, by rotating a mirror which is reflecting a bright light. Rotating lighthouses at night send out beams which can be seen travelling with great rapidity. A beam is not a "thing," because it is not tangible, and yet, for common sense, it preserves its identity while it rotates. But common sense is not shocked when the beam is broken up into a series of events. A purely visual view of matter makes it much easier to regard all material things as series of events, like the rotating beam.

Of course I am not suggesting that the other senses should be ignored as sources of knowledge concerning the physical world. What I am saying is that physics has tended, more and more, to interpret the information derived from the other senses by means of an imaginative picture derived from sight. Perhaps there are reasons for this; indeed, two suggest themselves, one physical and one physiological. Anticipating later discussions, we may say that fairly accurate perception is only possible when there is a causal chain, leading from the object to the sense-organ, which is to a considerable extent independent of what is to be found in the intermediate regions. Whether this is the case or not is a question for physics. Touch is confined to bodies with which the observer is in contact; smell and sound are not diffused very far. But light-waves travel with extraordinarily little modification through empty space, and without very great modification through a clear atmosphere. If we were to accept Professor Lewis's theory mentioned in Chapter XIII., we could say that a light-quantum travels unchanged from a star to a human eye. Even if this theory is not true, the mere fact that it can be seriously proposed illustrates the causal "purity" (if I may use such a word) of the passage of light from one body to another. This is the physical merit of sight as a source of knowledge concerning the external world.

The other merit is physiological. One kind of physical

stimulus is better than another, as a source of information, if less energy is required to produce a noticeable sensation, and smaller physical differences are required to produce noticeable differences of sensation. In both these respects, light is peculiarly excellent. The energy in the light from a just perceptible star is of the order of one quantum per cubic metre.* Very small differences of wave-length produce perceptible differences of colour, and stars are seen as separate even when the angle between the rays from them to the eye is very minute. In these respects, sight is markedly the best of the senses. It is therefore not surprising that physics has laid increasing stress upon visual data.

At the level of common sense, the most important merit of sight is that it makes us aware of objects at a distance. Sound and smell do this to some extent—smell, however, is much more important to certain species of animals than to us. But neither sound nor smell carry over great distances, and they do not enable us to locate their source at all accurately. If we accept the usual causal theory of perception—as I think we should—the proximate physical cause of the physiological occurrences leading to a visual perception is not something happening in the object which we say we see, but something happening at the surface of the eye. If this is to give us information about the distant object, it must be, in the main, causally determined by the object, without regard to anything intervening between the object and the eye. This is the physical merit of sight which we mentioned a moment ago. It has, of course, very distinct limitations. The colour of the light which reaches the eye will be different from that emitted by the object if there is intervening mist or coloured glass. The direction can be altered by a refracting medium. Mirrors deceive animals and young children. Then there are more subtle matters, such as the Doppler effect and aberration. But

* Jeans, *op. cit.*, p. 29.

after making all these allowances, sight remains supreme as a method of acquiring knowledge about distant objects.

In one respect, sight is defective—namely, in regard to distance. Some psychologists argue that depth can be, to a certain extent, perceived by sight alone, while others contend that it is wholly derived from other data. However that may be, it is certain that sight alone cannot judge any but very small distances. No one can distinguish between a hundred yards and a hundred miles by sight alone. Infants do not know at all, at first, which visual objects are within their grasp and which are not. For practical purposes, visual space has only two dimensions, even if this is not strictly correct in psychological theory. In practice, when we know the " real " size of a distant object, say a man or a cow, we can judge its distance by its apparent size.* But our initial experience of distance is derived from the amount of bodily movement required to establish contact. We may only have to stretch out an arm, we may have to lean the body, or we may have to walk for some time. An hour's walk is a natural measure of distance—in fact, it is a league. We cannot arrive at the common-sense idea of space without bringing in movement. And measurement with a measuring rod involves movement, if the distance to be measured is longer than the rod. Of course there is space in our own body, which is known without movement: we refer a headache to the head and a stomach-ache to the stomach. But this space is limited, and does not give spatial relations between our body and objects merely seen. To acquire a knowledge of these relations, bodily movement is indispensable. And this would never have been available for the purpose if there were not so many objects surrounding us which are motionless relatively to the earth.

* To show the depth of Dover cliff, Shakespeare says:

> " The crows and choughs that wing the midway air
> Show scarce so gross as beetles."

We can discover the distance of a house by walking to it, but not of a fox by the distance we have to gallop before reaching him.

Science cannot dispense wholly with postulates, but as it advances their number decreases. I mean by a postulate something not very different from a working hypothesis, except that it is more general: it is something which we assume without sufficient evidence, in the hope that, by its help, we shall be able to construct a theory which the facts will confirm. It is by no means essential to science to assume that its postulates are true always or necessarily; it is enough if they are often true. They ought to be so used that, when they are true, they yield verifiable theories, but, when they are not true, *no* theory can be framed which will fit the facts—until we find a way of working with different postulates.

The most important postulate of science is induction. This may be formulated in various ways, but, however formulated, it must yield the result that a correlation which has been found true in a number of cases, and has never been found false, has at least a certain assignable degree of probability of being always true. I propose to assume the validity of induction, not because I know of any conclusive grounds in its favour, but because it seems, in some form, essential to science and not deducible from anything very different from itself. I do not propose to discuss it, because the problem concerns empirical knowledge in general, not physics in particular; also because the subject is so complicated that a discussion is useless unless it is very lengthy. For the moment I must refer the reader to Mr Keynes and his critics.*

* *A Treatise on Probability*. By John Maynard Keynes. Macmillan, 1920.
Le Problème Logique de l'Induction. Par Jean Nicod. Paris, Alcan, 1924.
Review of the above by Braithwaite, *Mind*, 1925.
The Foundations of Probability. By R. H. Nisbet. *Mind*, January, 1926.

The other postulates which were at one time thought necessary have gradually been found to be superfluous. At one time, the indestructibility of matter would have been regarded as a postulate. Now, though electrons and protons are supposed to persist as a rule, it is seriously suggested that an electron and a proton may sometimes combine so as to annihilate each other; Eddington has advanced this as an important possible source of stellar energy.* It is true that, in this process, energy is supposed to be not destroyed; but the conservation of energy is no more than an empirical generalization, and is not thought to be strictly true.

Spatio-temporal continuity was, until lately, a postulate of science, but the quantum theory has called it in question without intellectual disaster. It *may* be true, but we cannot say that it *must* be.

The existence of causal laws perhaps deserves to rank as a postulate, or may perhaps be proved probable, on the existing evidence, if induction is assumed. Here our proviso is relevant, that a postulate need not be supposed to hold universally. We shall assume that there are causal laws, and try to discover them; but if none are found in a given region, that merely means that science cannot conquer that region. There are at present important regions of this kind. We do not know why a radio-active atom disintegrates at one moment rather than another, or why a planetary electron changes its orbit at one moment rather than another. We cannot be sure that these occurrences severally are governed by laws; but if they are not, science cannot deal with them individually, and is confined to statistical averages. Whether this will prove to be the case, we cannot yet say.

* *Nature*, May 1, 1926, supplement.

CHAPTER XVII

WHAT IS AN EMPIRICAL SCIENCE?

IT would be generally agreed that physics is an empirical science, as contrasted with logic and pure mathematics. I want, in this chapter, to define in what this difference consists.

We may observe, in the first place, that many philosophers in the past have denied the distinction. Thorough-going rationalists have believed that the facts which we regard as only discoverable by observation could really be deduced from logical and metaphysical principles; thorough-going empiricists have believed that the premisses of pure mathematics are obtained by induction from experience. Both views seem to me false, and are, I think, rarely held in the present day; nevertheless, it will be as well to examine the reasons for thinking that there is an epistemological distinction between pure mathematics and physics, before trying to discover its exact nature.

There is a traditional distinction between necessary and contingent propositions, and another between analytic and synthetic propositions. It was generally held before Kant that necessary propositions were the same as analytic propositions, and contingent propositions were the same as synthetic propositions. But even before Kant the two distinctions were different, even if they effected the same division of propositions. It was held that every proposition is necessary, assertoric, or possible, and that these are ultimate notions, comprised under the head of "modality." I do not think much can be made of modality, the plausibility of which seems to have come from confusing propositions with propositional functions. Propositions may, it is true, be divided in a way corresponding to what was meant by analytic and synthetic; this will be explained in a moment. But propositions which are not analytic can only be true or false; a true synthetic proposition cannot have a further property of being necessary, and a false synthetic proposition cannot have the property of being possible. Propositional functions, on the contrary, are of three kinds: those which are true for all values of the argument or arguments, those which are false for all values, and those which are true for some arguments and false for others. The first may be called necessary, the second impossible, the third possible. And these terms may be transferred to propositions when they are not known to be true on their own account, but what is known as to their truth or falsehood is deduced from knowledge of propositional functions. *E.g.* "it is possible that the next man I meet will be called John Smith" is a deduction from the fact that the propositional function "x is a man and is called John Smith" is possible—*i.e.* true for some values of x and false for others. Where, as in this instance, it is worth while to say that a *proposition* is possible, the fact rests upon our ignorance. With more knowledge, we should know who is the next man I shall meet, and then it would be certain that he is John Smith or certain that he is not John Smith. Possibility in this sense thus becomes assimilated to probability, and may count as any degree of probability other than 0 and 1. An "assertoric" proposition, similarly, was, I think, a confused notion applicable to a proposition known to be true but also known to be a value of a propositional function which is sometimes false—*e.g.* "John Smith is bald."

The distinction of analytic and synthetic is much more relevant to the difference between pure mathematics and physics. Traditionally, an "analytic" proposition was one whose contradictory was self-contradictory, or, what came to the same thing in Aristotelian logic, one which ascribed to a subject a predicate which was part of it—*e.g.* "white horses are horses." In practice, however, an analytic proposition was one whose truth could be known by means of logic alone. This meaning survives, and is still important, although we can no longer use the definition in terms of subject and predicate or that in terms of the law of contradiction. When Kant argued that " $7+5=12$ " is synthetic, he was using the subject-predicate definition, as his argument shows. But when we define an analytic proposition as one which can be deduced from logic alone, then " $7+5=12$ " is analytic. On the other hand, the proposition that the sum of the angles of a triangle is two right angles is synthetic. We must ask ourselves, therefore: What is the common quality of the propositions which can be deduced from the premisses of logic ?

The answer to this question given by Wittgenstein in his *Tractatus Logico-Philosophicus* seems to me the right one. Propositions which form part of logic, or can be proved by logic, are all *tautologies*—*i.e.* they show that certain different sets of symbols are different ways of saying the same thing, or that one set says part of what the other says. Suppose I say: " If p implies q, then not-q implies not-p." Wittgenstein asserts that " p implies q " and " not-q implies not-p " are merely different symbols for one proposition: the fact which makes one true (or false) is the same as the fact which makes the other true (or false). Such propositions, therefore, are really concerned with symbols. We can know their truth or falsehood without studying the outside world, because they are only concerned with symbolic manipulations. I should add—though here Wittgenstein might dissent—that all pure mathematics consists of tautologies in the above sense. If this is true, then obviously empiricists such as J. S. Mill are wrong when they say that we believe $2+2=4$ because we have found so many instances of its truth that we can make an induction by simple enumeration which has little chance of being wrong. Every unprejudiced person must agree that such a view *feels* wrong: our certainty concerning simple mathematical propositions does not seem analogous to our certainty that the sun will rise to-morrow. I do not mean that we feel more sure of the one than of the other, though perhaps we ought to do so; I mean that our assurance seems to have a different source.

I accept the view, therefore, that some propositions are tautologies and some are not, and I regard this as the distinction underlying the old distinction of analytic and synthetic propositions. It is obvious that a proposition which is a tautology is so in virtue of its form, and that any constants which it may contain can be turned into variables without impairing its tautological quality. We may take as a stock example: "If Socrates is a man and all men are mortal, then Socrates is mortal." This is a value of the general logical tautology:

"For all values of x, α, and β, if x is an α, and all α's are β's, then x is a β."

In logic, it is a waste of time to deal with particular examples of general tautologies; therefore constants ought never to occur, except such as are purely formal. The cardinal numbers turn out to be purely formal in this sense; therefore all the constants of pure mathematics are purely formal.

A proposition cannot be a tautology unless it is of a certain complexity, exceeding that of the simplest propositions. It is obvious that there is more complexity in equating two ways of saying the same thing than there is in either way separately. It is obvious also that, whenever it is actually useful to know that two sets of symbols say the same thing, or that one says part of what the other says, that must be because we have some knowledge as to the truth or falsehood of what is expressed by one of the sets. Consequently logical knowledge would be very unimportant if it stood alone; its importance arises through its combination with knowledge of propositions which are not purely logical.

All the propositions which are not tautologies we shall call "synthetic." The simplest kinds of propositions must be synthetic, in virtue of the above argument. And if logic or pure mathematics can ever be employed in a process leading to knowledge that is not tautological, there must be sources of knowledge other than logic and pure mathematics.

The distinctions hitherto considered in this chapter have been logical. In the case of modality, it is true, we found a certain confusion from an admixture of epistemological notions; but modality was intended to be logical, and in one form it was found to be so. We come now to a distinction which is essentially epistemological, that, namely, between *a priori* and empirical knowledge.

Knowledge is said to be *a priori* when it can be acquired without requiring any fact of experience as a premiss; in the contrary case, it is said to be empirical. A few words are necessary to make the distinction clear. There is a process by which we acquire knowledge of dated events at times closely contiguous to them; this is the process called "perception" or "introspection"* according to the character of the events concerned. There is no doubt need of much discussion as to the nature of this process, and of still more as to the nature of the knowledge to be derived from it; but there can be no doubt of the broad fact that we do acquire knowledge in this way. We wake up and find that it is daylight, or that it is still night; we hear a clock strike; we see a shooting star; we read the newspaper; and so on. In all these cases we acquire knowledge of events, and the time at which we acquire the knowledge is the same, or nearly the same, as that at which the events take place. I shall call this process "perception,"

* I do not wish to prejudice the question whether there is such a process as "introspection," but only to include it *if* it exists.

and shall, for convenience, include introspection—if this is really different from what is commonly called "perception." A fact of "experience" is one which we could not have known without the help of perception. But this is not quite clear until we have defined what we mean by "could not"; for clearly we may learn from experience that $2+2=4$, though we afterwards realize that the experience was not logically indispensable. In such cases, we see afterwards that the experience did not prove the proposition, but merely suggested it, and led to our finding the real proof. But, in view of the fact that the distinction between empirical and *a priori* is epistemological, not logical, it is obviously possible for a proposition to change from the one class to the other, since the classification involves reference to the organization of a particular person's knowledge at a particular time. So regarded, the distinction might seem unimportant; but it suggests some less subjective distinctions, which are what we really wish to consider.

Kant's philosophy started from the question: How are synthetic *a priori* judgments possible? Now we must first of all make a distinction. Kant is concerned with *knowledge*, not with mere *belief*. There is no philosophical problem in the fact that a man can have a *belief* which is synthetic and not based on experience—*e.g.* that this time the horse on which he has put his money will win. The philosophical problem arises only if there is a class of synthetic *a priori* beliefs which is always true. Kant considered the propositions of pure mathematics to be of this kind; but in this he was misled by the common opinion of his time, to the effect that geometry, though a branch of pure mathematics, gave information about actual space. Owing to non-Euclidean geometry, particularly as applied in the theory of relativity, we must now distinguish sharply between the geometry applicable to actual space, which is an empirical study forming part of physics, and the geometry of pure mathematics, which gives no information as to actual space. Consequently this instance of synthetic *a priori* knowledge, upon which Kant relied, is no longer available. Other kinds have been supposed to exist—for example, ethical knowledge, and the law of causality; but it is not necessary for our purposes to decide whether these kinds really exist or not. So far as physics is concerned, we may assume that all real knowledge is either dependent (at least in part) upon perception, or analytic in the sense in which pure mathematics is analytic. The Kantian synthetic *a priori* knowledge, whether it exists or not, seems not to be found in physics—unless, indeed, the principle of induction were to count as such.

But the principle of induction, as we have already seen, has its origin in physiology, and this suggests a quite different treatment of *a priori* beliefs from that of Kant. Whether there is *a priori knowledge* or not, there undoubtedly are, in a certain sense, *a priori beliefs*. We have reflexes which we intellectualize into beliefs; we blink, and this leads us to the belief that an object touching the eye will hurt it. We may have this belief before we have experience of its truth; if so, it is, in a sense, synthetic *a priori* knowledge—*i.e.* it is a belief, not based upon experience, in a true synthetic proposition. Our belief in induction is essentially analogous. But such beliefs, even when true, hardly deserve to be called knowledge, since they are not all true, and therefore all require verification before they ought to be regarded as certain. These beliefs have been useful in generating science, since they supplied hypotheses which were largely true; but they need not survive untested in modern science.

I shall therefore assume that, at any rate in every department relevant to physics, all knowledge is either analytic in the sense in which logic and pure mathematics are analytic, or is, at least in part, derived from perception. And all knowledge which is in any degree necessarily dependent upon perception I shall call "empirical." I shall regard a piece of knowledge as necessarily dependent upon perception when, after a careful analysis of our grounds for believing it, it is found that among these grounds there is the cognition of an event in time, arising at the same time as the event or very shortly after it, and fulfilling certain further criteria which are necessary in order to distinguish perception from certain kinds of error. These criteria will occupy us in the next chapter.

In a science, there are two kinds of empirical propositions. There are those concerned with particular matters of fact, and those concerned with laws induced from matters of fact. The appearances presented by the sun and moon and planets on certain occasions when they have been seen are particular matters of fact. The inference that the sun and moon and planets exist even when no one is observing them—in particular, that the sun exists at night and the planets by day—is an empirical induction. Heraclitus thought the sun was new every day, and there was no logical impossibility in this hypothesis. Thus empirical laws not only depend upon particular matters of fact, but are inferred from these by a process which falls short of logical demonstration. They differ from propositions of pure mathematics both through the nature of their premisses and through the method by which they are inferred from these premisses.

In an advanced science such as physics, the part played by pure mathematics consists in connecting various empirical generalizations with each other, so that the more general laws which replace them are based upon a larger number of matters of fact. The passage from Kepler's laws to the law of gravitation is the stock instance. Each of the three laws was based upon a certain set of facts; all three sets of facts together formed the basis of the law of gravitation. And, as usually happens in such cases, new facts, not belonging to any of the three previous sets, were found to support the new law—for instance, the facts of tides, of lunar motion, and of perturbations. Epistemologically, in such cases, a fact is a premiss for a law; logically, most of the relevant facts are consequences of the law—*i.e.* all except those required to determine the constants of integration.

In history and geography, the empirical facts are, at present, more important than any generalizations based upon them. In theoretical physics, the opposite is the case: the fact that the sun and moon exist is chiefly interesting as affording evidence of the law of gravitation and the laws of the transmission of light. In a philosophic analysis of physics, we need not consider particular facts except when they form the evidence for a theory. It is of course part of the business of such an analysis to consider what all particular facts have in common, and how they come to be known; but such inquiries are general. We are interested in the concept of topography, but not in the actual topography of the universe; at least, we are not interested in it for its own sake, but only as affording the evidence for general laws.

We have, in view of the above considerations, several different matters to consider, before we can return to actual physics. We have first to consider the nature and validity of the process we have called "perception"; next we have to investigate the general character of the facts known by perception; and lastly we have to examine the inference from facts of perception to empirical laws. After disposing of these topics, we shall resume contact with physics, asking ourselves now, not what physics asserts, but what justification it has for its assertions, and what inessential modifications will increase this justification.

CHAPTER XVIII

OUR KNOWLEDGE OF PARTICULAR MATTERS OF FACT

In this chapter, I wish to consider whatever would ordinarily pass for knowledge of particular matters of fact, in so far as this is not obtained by a process of deliberate scientific inference. I want to consider this as far as possible independently of the scientific laws based upon it, though not completely without reference to the primitive beliefs by which common sense draws inferences from perceptions. In particular, I wish to abstain from introducing the causal theory of perception, unless, on investigation, this should prove impossible. It will be understood that my purpose is epistemological: I am considering perception because it is involved in the premisses of empirical sciences, not because it is interesting as a mental process. It is of course necessary to consider its intrinsic character, but we do not do this for its own sake, we do it for the sake of the light that it may throw upon the character and extent of our knowledge.

We are met at the outset by a difficulty due to the fact that philosophical terminology is inappropriate when the views to be expressed are in any way unusual. "Knowledge" and "belief" both have connotations which are inconvenient for the purpose I have in view. They are both commonly applied in orthodox psychology to something conscious and explicit, such as is, or may be, already expressed in words. For our purposes, it is desirable to include more primitive occurrences, such as may be supposed to exist in animals. Obviously a bird can see an approaching man, and fly away in consequence. I wish to include under "perception" what happens in the bird, and also to say that the bird "knows" something when it sees a man, though I shall not venture to say what it knows.

But at this point a good deal of caution is necessary. My knowledge of the bird is part of my knowledge of the external world, and is partly, if not wholly, physical knowledge. Therefore when I am asking: how do *I* know about the physical world ? I have no right to begin by comparing my knowledge with that of a bird. I must start from myself and my own cognitions, and use the bird only to suggest hypotheses. This caution applies also to what was said in Chapter XV.

Again, there is always a danger, in epistemology, of putting the less certain before the more certain. My knowledge of the process of perceiving is less certain, and less primitive, than my knowledge of percepts. When I say, " I know that I have just heard a clap of thunder," I am saying something not so indubitable as when I say, " There has just been a clap of thunder." It is facts of this latter kind that are required as premisses in physics. A man might be completely competent as a physicist if he knew such propositions as " There has just been a clap of thunder " even if he knew no propositions such as " I know that there has just been a clap of thunder." The consideration of our knowing, as opposed to what we know, is forced on us by the fact that what we think we know sometimes turns out to be false; if this were not the case, an analysis of matter need not consider our knowing at all. As it is the case, we are compelled to examine our knowing, as well as what we know, with a view to discovering, if possible, how to minimize the risk involved in taking as knowledge what, on reflection, we still believe to be knowledge.

We are often urged to adopt an artificial naiveté in investigating problems concerning what we know; if we do not do so, we are accused of the " psychologist's fallacy." Now in certain problems this caution is quite proper, but in others it is not. My problem is: What do I, here and now, know about the external world, and how do I know it ? It is obvious that my knowledge of the external world cannot be dependent upon (say) how long it takes a fish to learn to recognize the man who feeds it, since this supposes that I know all about the fish and the man and the feeding. Facts about the perceptions of babies, such as we considered in Chapter XV., come under the same head. Long before I can know that there are babies, I must know many other things about the external world. I want to start from what comes epistemologically first in *my* existing knowledge *now*; and in this problem, obviously, I cannot assume that I already know all about the experiences of animals and babies. There must therefore be no artificial naiveté, but a straightforward investigation of my knowledge as I find it.

The position may be illustrated by Chuang-Tze's story of the two philosophers on the bridge. The first says: "See how the little fishes are darting about. Therein consists the pleasure of fishes." The second replies: "How do you, not being a fish, know wherein consists the pleasure of fishes?" To which the first retorts: "How do you, not being I, know that I do not know wherein consists the pleasure of fishes?" My position is that of the second philosopher. If other philosophers know "wherein consists the pleasure of fishes," I congratulate them; but I am not thus gifted.

When I try to disentangle the primitive from the inferred elements in what I take to be my knowledge, I find that the task is not really very difficult, except in certain niceties. The primitive part seems something like this: There are coloured shapes which move, there are noises, smells, bodily sensations, the experiences which we describe as those of touch, and so on. There are relations among these items: time-relations (earlier and later) among all of them, and space-relations (up-and-down, right-and-left, and the relations by which localization in the body is effected) among many of them. There are recollections of some of these things; this seems indubitable, although it is not easy to say in what a recollection consists, or how it is related to what it recollects. There are also expectations; by this I mean something just as immediate as memory. Everyone knows the story of the Orangeman who fell off a scaffolding and murmured as he fell: " To Hell with the Pope, and now for the—bump." He was experiencing expectation in the sense in which I mean it. Of thoughts other than memories and expectations, it is not necessary to take account when our sole purpose is to reach the primitive basis of our knowledge of matter.

In the above account, I have omitted many things which I formerly " knew," and which, apparently, most other people " know." I have omitted " objects." In former days, my apparatus of non-inferential knowledge included tables and chairs and books and persons and the sun and moon and stars. I have come to regard these things as inferences. I do not mean that I inferred them formerly, or that other people do so now. I fully concede that I did not infer them. But now, as the result of an argument, I have become unable to accept the knowledge of them as valid knowledge, except in so far as it can be inferred from such knowledge as I still consider epistemologically primitive.

The argument in question would naturally, but not validly, express itself in terms of the causal theory of perception. What I see—so it might be urged—is causally dependent upon the light waves that reach my eye, and these waves might be reflected or refracted in such a way as to deceive me concerning their source. This way of stating the argument is invalid because it assumes more knowledge of the physical world than we have any right to assume at our present level. But the facts upon which it relies can be easily made available, without any undue assumption of knowledge, for the purpose of proving our conclusion. In certain cases in which we seem to have immediate knowledge of objects, we find ourselves surprised by something totally unexpected. The dog listening to " his master's voice " on the gramophone may serve as an illustration. He thinks he perceives his master, but in fact he only perceives a noise. In restaurants which wish to look larger than they are, one whole wall sometimes consists of looking-glass, and it is easy to suppose that one perceives diners at tables, when in fact they are mere reflections. Perspective can be made to deceive. When I say " deceive," in this connection, I mean " rouse expectations which are not fulfilled." It is useless to multiply examples. The upshot is that what seems like perception of an object is really perception of certain sensible qualities together with expectations of other sensible qualities—the commonest case being something visual which rouses tactual expectations. It is found that the occasional deceptive experiences are not, in themselves, distinguishable from those that are not deceptive. Hence we conclude that we have to do with a correlation which is usual but not invariable, and that, if we wish to construct an exact science, we must be sceptical of the associations which experience has led us to form, connecting sensible qualities with others with which they are often but not always combined.

The above argument is based upon principles which common sense can be brought to accept, and has a conclusion which physics has accepted, though perhaps without fully realizing its scope. The argument is not " philosophical," in the sense of coming from a region quite different from that of science and ordinary knowledge. It proceeds merely on the usual principle of trying to substitute something more accurate for a belief which has been found to lead to error on occasion. It has as a consequence that " matter," in physics and in philosophy, if legitimate at all, cannot be altogether identified with the common-sense notion of a material object, though it will have a certain connection with this notion, since the common-sense belief in material objects does not *usually* lead to false expectations.

Some misunderstandings must be guarded against as regards expectation and error. Neither of these is primarily intellectual; I should be inclined to say that both are primarily muscular—or, we may say, nervous, in order not to seem paradoxical. Suppose you set to work to lift a watering-can: you may adjust your muscles in the way appropriate if the can is full, or in the way appropriate if it is empty. If they are adjusted to a full can when the can is empty, you receive a shock of surprise on experiencing the lightness of the can. You would describe your experience by saying, " I thought the can was full of water." But as a rule, in such situations, there has not been anything that could be called " thought "; there has been physiological adjustment as a result of a stimulus. Of course there *may* have been " thought "; and whatever " thought " may be, it certainly can produce the kind of muscular effects which we are considering. But these effects can be produced more directly, and usually are. There is so little essential difference between a process involving "thought" and one not involving it that it seems a mistake to confine the notions of truth and error to intellectual processes; they ought rather, it seems to me, to be applied to the complete reaction of a person to a situation, in which " thought " is only one element. But it will not do, at our present level, to introduce physiology, since we are considering how we know about matter, and must not therefore assume that we already know about the matter in our own body. However, the phenomena are easily described in the way which our problem demands. In the case of the watering-can, the vivid part of the experience is the surprise. But by means of attention a number of other elements can be observed. We can observe the feelings which are interpreted as meaning muscular adjustment to a heavy load; we can observe the visual appearance described as the can coming up with a jerk; we can observe the sudden change in what, for short, we may call muscular feelings. It is impossible to describe all this without circumlocution, since the natural words to use presuppose physiology; but it is clear that there is a great deal that can be directly observed, without invoking any theory. In such a process, what comes earlier may be described as "error" because of the emotion of surprise which follows. Where the activity which has been begun runs its course without leading to this emotion, we shall say that there is not error. I hesitate to ascribe "truth" to something pre-intellectual, but at any rate we may say that there is "correctness," or that what has succeeded to the sensation (or perception) which came at the beginning of the process has been "correct." We may shorten this by saying that the response to a stimulus may be "correct" or "erroneous." But the longer phrase has the merit of not assuming so much knowledge of causal relations.

In the situations to which the above analysis applies, we have the advantage of a perfectly definite criterion of correctness or error. The feeling of surprise marks error, and the absence of this feeling marks correctness. It must not be supposed that we have normally an explicit prevision, still less an explicit inference; all that can be said is that we are in such a condition that one sort of event will cause surprise while another sort will not. Consider the experience we have all had, of "thinking" we were at the bottom of a staircase when in fact there was another step to go down. In such a case, when we "think" we are at the bottom, we do not think at all, for if we did we should not make such a silly mistake. Indeed, we might say (or an Irishman might): "I thought I was at the bottom because I wasn't thinking."

It is fairly clear that all our elementary intellectual processes have pre-intellectual analogues. The analogue of a general causal belief is a reflex or a habit. A dog goes to the dining-room when he hears the dinner-bell, and so do we. In the case of the dog, it is easy to suppose that he has merely acquired a habit, without having formulated the induction: "Dinner-bells are a cause, or an effect, or an indispensable part of the cause, of dinner." We, however, can formulate this induction, and we shall then suppose that it is because we have done so that we go into the dining-room when we hear the bell. In fact, however, we may be just as merely habitual as the dog. The elementary inductions of common sense are first habits, and only subsequently beliefs. We may say that if, in our experience, *A* is accompanied by *B* either often or in some emotionally important manner, this fact causes first a habit which would be rational if *A* were always accompanied by *B*, and then a belief that *A* is always accompanied by *B*—the latter being a rationalization of the pre-existing habit.

General propositions may thus form part of our thinking from the start. Such general propositions are merely the verbal expression of habits. The hand-eye co-ordination becomes firmly fixed as a motor habit, and then, when we think, we conclude that what can be seen can often be touched—in fact, that it can be touched in circumstances which we know in practice, though we might have difficulty in formulating them exactly. Such general propositions are synthetic, and are in a certain sense *a priori*; for, though experience has *caused* them, they are not obtained by inference from other propositions, but by rationalizing and verbalizing our habits; that is to say, their antecedents are pre-intellectual. The trouble with them is that they are never quite right. Common sense, do what it will, cannot avoid being surprised occasionally. The object of science is to spare it this emotion, and create mental habits which shall be in such close accord with the habits of the world as to secure that nothing shall be unexpected. Science has, of course, not yet achieved its ideal: the Great War and the earthquake of Tokyo took people by surprise. But it is hoped that in time such events will no longer disturb us, because we shall have expected them. However, I do not wish at this stage to consider our knowledge of general propositions; it is particular matters of fact that concern us at present.

Although, in our less intellectual moods, we act as the result of a sensation without stopping to think (*e.g.* when we blink because we see something approaching the eye), yet we can, when we choose, react to a stimulus in the way which is called "knowing" it, and we often react involuntarily in this way. It is not necessary, in an analysis of matter, to decide what "knowing" is; it is only necessary to decide what is known, in so far as this is relevant to our knowledge of physics. The list which I gave earlier in the present chapter was designed to be such as would exclude the risk of error, using "error" in the sense which I have been defining. Common sense is liable to err—of this we have already given instances. We cannot therefore include the common-sense notion of an "object" or "thing" as part of what we know. But the sensible qualities which can be analyzed out of the "thing" can be admitted without ever leading us into error. These, therefore, are to be accepted as genuinely known.

It is a remarkable fact that all such knowledge, when not inferential, arises at about the same time as what is known, though it may survive for an indefinite time in the form of memory. This is the essential peculiarity, which we mentioned earlier, that distinguishes the empirical premisses of empirical knowledge. These consist of facts which become known spontaneously at about the time when they occur, and cannot be known sooner except by elaborate and more or less doubtful inferences from other such facts. The process of getting to know such facts without inference is called "perception," and knowledge derived wholly or partly from perception is said to be based on experience. A Greek could know the multiplication table as well as we do, but he could not know the biography of Napoleon.

CHAPTER XIX

DATA, INFERENCES, HYPOTHESES, AND THEORIES

WHEN a man of science speaks of his "data," he knows very well in practice what he means. Certain experiments have been conducted, and have yielded certain observed results, which have been recorded. But when we try to define a "datum" theoretically, the task is not altogether easy. A datum, obviously, must be a fact known by perception. But it is very difficult to arrive at a fact in which there is no element of inference, and yet it would seem improper to call something a "datum" if it involved inference as well as observation. This constitutes a problem which must be briefly considered.

What is recorded as the result of an experiment or observation is never the bare fact perceived, but this fact as interpreted by the help of a certain amount of theory. Take, say, the eclipse observations by which Einstein's theory of gravitation was confirmed. What in fact was given in perception was—apart from the previous arrangements—a visual pattern of dots, interpreted as a photograph of stars near the sun; a tactual-visual experience called "measuring," and finally coincidences of certain visual appearances with certain others called "numbers on a scale." At least, whether this is actually a correct account or not, it represents the sort of thing that occurred. A considerable amount of theory was involved in merely measuring the photographs. And of course a vast structure was involved in interpreting the photographs as photographs of stars, and in inferring thence the course which the light from the stars had pursued. It is the theoretical element in measuring the photographs that most needs to be stressed, since it is easily overlooked.

It is sometimes maintained that there is something of the nature of inference at an even earlier stage. The effects of a given sensory stimulus upon two men with indistinguishable sense-organs but different experiences may be very different. The most obvious illustration is the effect of print upon a man who can read and upon a man who cannot. A child learning to read is aware of each letter in turn as a certain shape, and finally arrives, with pain and labour, at the word. A man who learned to read as a child is quite unconscious of the letters, unless he is interested in typography or looking out for misprints; normally, he passes straight to the words, and to the words as having meaning, not as black marks on white paper. Nevertheless, he is very likely to notice an oddity at once—say if someone omitted the *z* in " Nietzsche." In writing to a philosopher to ask for a testimonial, it would be very unsafe to assume that he would not detect an error of this sort. But the detection of the error is due to the element of surprise: the philosopher is expecting a *z*, and has a shock when it is not there, like that of a man who has reached the bottom of a staircase but thinks there is another step. The philosopher's body was expecting a *z*, though his mind was otherwise occupied.

A more orthodox illustration is the difference between the effect of a visual stimulus upon an ordinary man and upon a man born blind but enabled to see as the result of an operation. The latter has not the tactual associations of the ordinary man, and cannot " interpret " what he sees. Are we to include in perception this element of unconscious interpretation, or are we to include only what we imagine that the same stimulus would have produced if there had been no such previous experience as would make interpretation possible ? This is not an altogether easy question. On the one hand, the interpretation depends upon correlations which are frequent but probably not invariable, so that, if it is included, it might seem as though perception would sometimes contain an element of error. On the other hand, the element of interpretation can only be eliminated by an elaborate theory, so that what remains—the hypothetical bare " sensation "—is hardly to be called a " datum," since it is an inference from what actually occurs. This last argument is, to my mind, conclusive. Perception must include those elements which are irreducibly physiological, but it need not on that account include those elements which come, or can be made to come, within the sphere of conscious inference. When we hear (say) a donkey braying, we are quite conscious of inference from the noise of the donkey, or at any rate we can easily become conscious of it. I should not, therefore, in this case, include anything else of the donkey with the perception, but only the noise. And if you see a donkey, though you may have reactions connected with the sense of touch, these are never confounded with what you feel when you actually touch him. I should therefore say that a great deal of the interpretation that usually accompanies a perception can be made conscious by mere attention, and that this part ought not to be included in the perception. But the part which can only be discovered by careful theory, and can never be made introspectively obvious, ought to be included in the perception. Perhaps the line between the two is not so sharp as could be wished; but I do not see how else to meet the conflicting considerations which present themselves.

We have still to ask ourselves whether perception, so defined, will sometimes contain an element of error. Here we must distinguish. It may be, and often is, accompanied by expectations which are disappointed; and we agreed to take this as the mark of error. But the expectations can be distinguished from the perception, although in practice this may not always be easy. The tactual accompaniments of visual perceptions are of the nature of expectations. There are no such accompaniments of perceptions of the heavenly bodies. I think that in all cases in which error occurs it is easy to distinguish the erroneous expectation from the perception. Whatever "interpretation" does not involve expectations need not be regarded as erroneous. It is supposed that indistinguishable stimuli may fall upon indistinguishable sense-organs, and yet result in distinguishable perceptions because of differences in the brains of the two percipients—these differences in their brains being the result of different experiences. But there is not on that account anything erroneous in the perception of either. A different event occurs in the one from that which occurs in the other; but each event really occurs. This topic, however, cannot be adequately discussed until we come to the causal theory of perception and the relation between perception and physical stimulus.

I come now to the question of inferences, which has already been touched on. As we have seen, there is a purely physiological form of inference which belongs to an earlier stage than explicit inference, though it persists in the habits of even the most sophisticated philosopher, such as Hume. The next stage is where there is an actual passage from one belief to another, but the passage is a mere occurrence, not a transition motived by an argument. In this case, the transition is usually caused by a physiological inference. Then there is inference based upon some belief; but even then the belief may be wholly irrational, or it may not logically warrant the inference, which is the case of fallacious reasoning. Lastly, there is valid inference by means of a true principle—but of this I cannot give an indubitable instance.

In historical fact, these types of inference emerge successively, but a later type does not cause an earlier one to disappear. Moreover, the later type tends to be adapted to the earlier. First we have physiological inference: this is exemplified when a bird flies so as not to bump into solid objects, and fails when it bumps into a window-pane. Then there is the transition from the belief expressing the premiss of the physiological inference to that expressing its conclusion, without any consciousness of how the transition is effected. Then there is belief in a causal law which is the intellectualized expression of the habit embodied in the physiological inference. And last of all there is the search for criteria by which to distinguish between true and false causal laws, these criteria being intellectual, not mere habits of the body. This last stage is only reached when we come to science.

One of the main purposes of scientific inference is to justify beliefs which we entertain already; but as a rule they are justified with a difference. Our pre-scientific general beliefs are hardly ever without exceptions; in science, a law with exceptions can only be tolerated as a makeshift. Scientific laws, when we have reason to think them accurate, are different in form from the common-sense rules which have exceptions: they are always, at least in physics, either differential equations, or statistical averages. It might be thought that a statistical average is not very different from a rule with exceptions, but this would be a mistake. Statistics, ideally, are accurate laws about large groups; they differ from other laws only in being about groups, not about individuals. Statistical laws are inferred by induction from particular statistics, just as other laws are inferred from particular single occurrences. All this, however, is by the way; the point is that inference as a practice has a long history before it becomes scientific.

The most important inference which science takes over from common sense is inference to unperceived entities. One form in which common sense makes this inference is that of a belief that objects which have been perceived still exist when they are not perceived. If, at a dinner-party, the electric light suddenly goes out, no one doubts that his neighbours and the dinner-table and the food and drink still exist, although at the moment they are unperceived. When the light goes on again, this belief appears to be confirmed; if there are fewer spoons than before, we do not infer that they have ceased to exist, but that someone present is a thief. This belief in the permanence of perceived objects has gone through all stages from physiological inference to advanced scientific or philosophical theory; the inquiry into its justification is the central problem in the analysis of matter, philosophically considered. No one, not even Berkeley, has treated it with quite the seriousness that it deserves, because the physiological inference is so irresistible that it is difficult to achieve a purely intellectual attitude towards the problem. This inference is the source of the philosophical notion of "substance" and the physical notion of "matter." For the present, I am only noting the inferences to be considered; I am not attempting to investigate their validity.

Unperceived entities are also inferred by common sense when it believes that other people have "minds." I wish to make it clear that even the most rigid behaviourist makes this inference, although in a slightly different form. Dr Watson, for example, would admit that his own toothache can lead him to say, "I have a toothache," whereas another person's toothache will not lead him to say "You have a toothache" without some intermediate link. Whatever may be our analysis of "knowledge," we certainly know things about our own bodies in ways which are not open to us where other people's bodies are concerned. There is nothing mysterious about this: it is analogous to the fact that some sounds are within earshot while others are not. The point is that we infer, from the behaviour of others, the existence of things (such as toothaches) which we cannot perceive. Whether we say that these things are "mental" or "bodily" makes no difference to the fact that we make inferences. These inferences, also, are at first purely physiological.

From the point of view of physics, the inference to other people's "minds" has a twofold importance. The first, which is not specially physical, is concerned with testimony. What is commonly accepted as the experimental evidence on any topic of physics includes not only what a given physicist has himself observed, but whatever has been reliably recorded. Everything that we learn from what other people say and write involves inference from something perceived (spoken or written words) to something unperceived—namely, the "mental" events of the speaker or writer. It may be that the primary inference is only to another person's percepts, but it is none the less an inference to something which *we* do not perceive. The second point about the inference to other people's percepts is specially physical; it concerns the fact that different people live in a common world. The percepts of two different people, if we accept testimony, are found to be often very similar, though not exactly alike; this leads to the theory of a common external cause—*i.e.* to the causal theory of perception, and to the division of the qualities of the perceived object into such as belong to the external cause and such as are supplied by the body or mind of the percipient.

The development of science out of common sense has not been by way of a radically new start at any moment, but rather by way of successive approximations. That is to say, where some difficulty has arisen which current common sense could not solve, a modification has been made at some point, while the rest of the common-sense view of the world has been retained. Subsequently, using this modification, another modification has been introduced elsewhere; and so on. Thus science has been an historical growth, and has assumed, at each moment, a more or less vague background of theory derived from common sense. This is one difference between science and philosophy: philosophy attempts, though not always successfully, to set out its inferences in a form which assumes nothing on the mere ground that it has always been assumed hitherto. It may be doubted whether science can retain its vitality if it is severed from its root in our animal habits; when set forth quite abstractly, it loses plausibility. Induction, for example, is difficult to justify, and yet indispensable in science. In such cases, I shall allow myself to accept what seems necessary on pragmatic grounds, being content, as science is, if the results obtained are often verifiably true and never verifiably false. But wherever a principle is accepted on such grounds as these, the fact should be noted, and we should realize that there remains an intellectual problem, whether soluble or not.

The actual procedure of science consists of an alternation of observation, hypothesis, experiment, and theory. The only difference between a hypothesis and a theory is subjective: the investigator believes the theory, whereas he only thinks the hypothesis sufficiently plausible to be worth testing. A hypothesis should accord with all known relevant observations, and suggest experiments (or observations) which will have one result if the hypothesis is true, and another if it is false. This is an ideal: in actual fact, other hypotheses will always exist which are compatible with what is meant to be an *experimentum crucis*. The crucial character can only be as between *two* hypotheses, not as between one hypothesis and all the rest. When a hypothesis has passed a sufficient number of experimental tests, it becomes a theory. The argument in favour of a theory is always the formally invalid argument: "p implies q, and q is true, therefore p is true." Here p is the theory, and q is the observed relevant facts. We are most impressed when q is very improbable *a priori*. For example,*

* Sommerfeld, *op. cit.*, p. 217.

observation gives Rydberg's constant as:

$$R = 1{\cdot}09678.10^5\ cm^{-1},$$

while Bohr's theory gives:

$$R = 1{\cdot}09.10^5\ cm^{-1},$$

which is within the degree of accuracy to be expected if the theory is right. Numerical confirmations of this kind are always the most striking. Nevertheless, even they must be received with caution; Bohr's theory of circular orbits required modification by the admission of elliptic orbits, and thus turned out to be not the only theory which would give a correct value of Rydberg's constant.

When a theory fits a number of facts, but goes slightly astray in regard to certain others, it happens generally, though not always, that it can be absorbed, by a slight modification, into a new theory which includes the hitherto discrepant facts. There are exceptions, of which the theory of relativity is perhaps the most notable: here an immense theoretical reconstruction was required to account for very minute discrepancies. But in general a partially successful theory is an essential step towards its successor. And a result deduced from a hitherto successful theory is more likely to be right than the theory is: the theory is only right if *all* its consequences are true (at least, so far as they can be tested), but a verifiable consequence of the theory is likely to be true if *most* of the verifiable consequences are true. That is why the practical value of scientific theories is so much greater than their philosophic value as contributions to ultimate truth. To some extent, we can distinguish, among the consequences of a theory, which are the most reliable; they will be those in the region of the facts which have given rise to the theory. No one is surprised to find that an empirical law connecting specific heat with temperature fails for temperatures much lower than those for which it has been found to be correct; but if, in the middle of these latter, there was found to be a small range of temperatures where the law failed, we should be very much surprised. Thus there is a kind of common sense to be used in applying theories: some applications can be made with confidence, while others will be felt to be questionable.

CHAPTER XX

THE CAUSAL THEORY OF PERCEPTION*

COMMON sense holds—though not very explicitly—that perception reveals external objects to us directly: when we " see the sun," it is the sun that we see. Science has adopted a different view, though without always realizing its implications. Science holds that, when we " see the sun," there is a process, starting from the sun, traversing the space between the sun and the eye, changing its character when it reaches the eye, changing its character again in the optic nerve and the brain, and finally producing the event which we call " seeing the sun." Our knowledge of the sun thus becomes inferential; our direct knowledge is of an event which is, in some sense, " in us." This theory has two parts. First, there is the rejection of the view that perception gives direct knowledge of external objects; secondly, there is the assertion that it has external causes as to which something can be inferred from it. The first of these tends towards scepticism; the second tends in the opposite direction. The first appears as certain as anything in science can hope to be; the second, on the contrary, depends upon postulates which have little more than a pragmatic justification. It has, however, all the merits of a good scientific theory—*i.e.* its verifiable consequences are never found to be false. Epistemologically, physics might be expected to collapse if perceptions have no external causes; therefore the matter must be examined before we can go further.

We must first give somewhat more precision to the common-

* On this subject, *cf.* chap. iv. of Dr Broad's *Perception, Physics, and Reality*, Cambridge, 1914.

sense view which is rejected by the causal theory. We have to ask what is meant by "external objects." One would naturally say "spatially external." But "space" is very ambiguous: in visual space, the objects which we see are mutually external, and objects other than the visual appearances of parts of our own body are spatially external to those appearances. In the space derived from the combination of touch and sight and bodily movement, which is the ordinary space of common sense, there is the same externality of visual appearances other than those of parts of our own body. Thus spatial externality, in the sense in which space can be derived from the relations of our own percepts, is not what is meant. I think we shall come nearer to what is meant if we say that two people can perceive the same object. In some sense, unless we reject testimony, we must of course admit that this is true: we can all see the sun unless we are blind. But this fact is differently interpreted by common sense and by the causal theory: for common sense, the percepts are identical when two people see the sun, whereas for the causal theory they are only similar and related by a common causal origin.

It would be a waste of time to recapitulate the arguments against the common-sense view. They are numerous and obvious and generally admitted. The laws of perspective may serve as an illustration: where one man sees a circle, another sees an ellipse, and so on. These differences are not due to anything "mental," since they appear equally in photographs from different points of view. Common sense thus becomes involved in contradictions. These do not exist for solipsism, but that is a desperate remedy. The alternative is the causal theory of perception.

We must not expect to find a *demonstration* that perceptions have external causes, which may produce perceptions in a number of people at the same time. The most that we can hope for is the usual ground for accepting a scientific theory—namely, that it links together a number of known facts, that it does not have any demonstrably false consequences, and that it sometimes enables us to make predictions which are subsequently verified. All these tests the causal theory fulfils; it must not be assumed, however, that no other theory could fulfil them. But let us examine the evidence.

First: there can be no question of logical proof. A certain collection of facts is known to me by perception and recollection; what else I believe about the physical world is either the effect of unreasoning habit or the conclusion of an inference. Now there cannot be any logical impossibility in a world consisting of just that medley of events which I perceive or remember, and nothing else. Such a world would be fragmentary, absurd, and lawless, but not self-contradictory.* I am aware that, according to many philosophers, such a world would be self-contradictory. I am aware also that, according to other philosophers, what we perceive is not fragmentary, but really embraces the whole universe—what is fragmentary is only what we perceive that we perceive. The first of these views is that of Hegel and his followers; the second is that of Bergson and (perhaps) of Dr Whitehead. The Hegelian view rests upon an elaborate logic, which I have controverted on former occasions; at present I am content to refer to what I have written before. The other view is traditionally associated with mysticism; my reasons for not accepting it are given in *Mysticism and Logic*. I say, therefore, on grounds given in former writings, that the world of perception and memory is fragmentary, but not self-contradictory. On grounds of logic, I hold that nothing existent can imply any other existent except a part of itself, if implication is taken in the sense of what Professor G. I. Lewis calls "strict implication," which is

* Perhaps it would not really be lawless; I shall discuss this at a later stage.

the relevant sense for our present discussion. If this is true, it follows that any selection of the things in the world might be absent, so far as self-contradiction is concerned. Given a world consisting of particulars x, y, z, ... interrelated in various ways, the world which results from the obliteration of x must be logically possible. It follows that the world consisting only of what we perceive and recollect cannot be self-contradictory; if, therefore, we are to believe in the existence of things which we neither perceive nor recollect, it must be either on the ground that we have other non-inferential ways of knowing matters of fact, or on the basis of an argument which has not the type of cogency that we should demand in pure mathematics, in the sense that the conclusion is only probable. As for the fragmentary character of the perceived world, those who deny it have to introduce minute perceptions, like Leibniz, or unconscious perceptions, or vague perceptions, or something of the kind. Now it seems to me unnecessary to inquire whether there are perceptions of such kinds; I certainly am not prepared to deny them dogmatically. But I do say that, even if they exist, they are useless as a basis for physics. Perceptions of which we are not sufficiently conscious to express them in words are scientifically negligible as data; our premisses must be facts which we have explicitly noted. Vagueness, no doubt, is omnipresent and unavoidable; but it is only in proportion as we overcome it that exact science becomes possible. And we overcome it most by analysis and concentration, not by a diffused ecstatic mystical vision.

I return now to the question: What grounds have we for inferring that our percepts and what we recollect do not constitute the entire universe? I believe that at bottom our main ground is the desire to believe in simple causal laws. But proximately there are other arguments. When we speak to people, they behave more or less as we should if we heard such words, not as we do when we speak them. When I say that they behave in a similar manner, I mean that our perceptions of their bodies change in the same sort of way as our perceptions of our own bodies would in correlative circumstances. When an officer who has risen from the ranks gives the word of command, he sees his men doing what he used to do when he heard the same sounds as a private; it is therefore natural to suppose that they have heard the word of command. One may see a crowd of jackdaws in a newly-ploughed field all fly away at the moment when one hears a shot; again it is natural to suppose that the jackdaws heard the shot. Again: reading a book is a very different experience from composing one; yet, if I were a solipsist, I should have to suppose that I had composed the works of Shakespeare and Newton and Einstein, since they have entered into my experience. Seeing how much better they are than my own books, and how much less labour they have cost me, I have been foolish to spend so much time composing with the pen rather than with the eye. All this, however, would perhaps be the better for being set forth formally.

First, there is a preliminary labour of regularizing our own percepts. I spoke of seeing others do what we should do in similar circumstances; but the similarity is obvious only as a result of interpretation. We cannot see our face (except the nose, by squinting) or our head or our back; but tactually they are continuous with what we can see, so that we easily imagine what a movement of an invisible part of our body ought to look like. When we see another person frowning, we can imitate him; and I do not think the habit of seeing ourselves in the glass is indispensable for this. But probably this is explained by imitative impulses—*i.e.* when we see a bodily action, we tend to perform the same action, in virtue of a physiological mechanism. This of course is most noticeable in children. Thus we first do what someone else has done, and then realize that what we have done is what he did. However, this complication need not be pursued. What I am concerned with is the passage, by experience, from "apparent" shapes and motions to "real" shapes and motions. This process lies within the perceptual world: it is a process of becoming acquainted with congruent groups—*i.e.* to speak crudely, with groups of visual sensations which correspond to similar tactual sensations. All this has to be done before the analogy between the acts of others and our own acts becomes obvious. But as it lies within the perceptual world, we may take it for granted. The whole of it belongs to early infancy. As soon as it is completed, there is no difficulty in interpreting the analogy between what we perceive of others and what we perceive of ourselves.

The analogy is of two kinds. The simpler kind is when others do practically the same thing as we are doing—for instance, applaud when the curtain goes down, or say "Oh" when a rocket bursts. In such cases, we have a sharp stimulus, followed by a very definite act, and our perception of our own act is closely similar to a number of other perceptions which we have at the same time. These, moreover, are all associated with perceptions very like those which we call perceptions of our own bodies. We infer that all the other people have had perceptions analogous to that of the stimulus to our own act. The analogy is very good; the only question is: Why should not the very same event which was the cause of our own act have been the cause of the acts of the others? Why should we suppose that there had to be a separate seeing of the fall of the curtain for each spectator, and not only one seeing which caused all the appearances of bodies to appear to applaud? It may be said that this view is far-fetched. But I doubt if it would be unreasonable but for the second kind of analogy, which is incapable of a similar explanation.

In the second kind of analogy, we see others acting as we should act in response to a certain kind of stimulus which, however, we are not experiencing at the moment. Suppose, for example, that you are a rather short person in a crowd watching election returns being exhibited on a screen. You hear a burst of cheering, but can see nothing. By great efforts, you manage to perceive a very notable result which you could not perceive a few moments earlier. It is natural to suppose that the others cheered because they saw this result. In this case, their perceptions, if they occurred, were certainly not *identical* with yours, since they occurred earlier; hence, if the stimulus to their cheering was a perception analogous to your subsequent perception, they had perceptions which you could not perceive. I have chosen a rather extreme example, but the same kind of thing occurs constantly; someone says "There's Jones," and you look round and see Jones. It would seem odd to suppose that the words you heard were not caused by a perception analogous to what you had when you looked round. Or your friend says "Listen," and after he has said it you hear distant thunder. Such experiences lead irresistibly to the conclusion that the percepts you call other people are associated with percepts which you do not have, but which are like those you would have if you were in their place. The same principle is involved in the assumption that the words you hear express "thoughts."

The argument in favour of the view that there are percepts, connected with other people, which are not among our own percepts, is presupposed in the acceptance of testimony, and comes first in logical order when we are trying to establish the existence of things other than our own percepts, both because of its inherent strength, and because of the usefulness of testimony in the further stages. The argument for other people's percepts seems to common sense so obvious and compelling that it is difficult to make oneself examine it with the necessary detachment. Nevertheless it is important to do so. As we have seen, there are three stages. The first does not take us outside our own percepts, but consists merely in the arrangement of them in groups. One group consists of all the percepts which common sense believes to be those of an identical object by different senses and from different points of view. When we eliminate reference to an object, a group must be constituted by correlations, partly between one percept and another (touch and sight when an object is held in the hand), partly between one percept and the changes in another (bodily movement and changes of visual and tactual perceptions while we move). In assuming that these correlations will hold in untested cases, we are of course using induction; otherwise, the whole process is straightforward. The process enables us to speak of a " physical object " as a group of percepts, and to explain what we mean by saying that a near object and a distant object are " really " of the same size and shape. Also we can explain what we mean by saying that a physical object does not " really " change as we walk away from it (*i.e.* as we have the percepts which make us say we are walking). This is the first stage in the argument.

In the second stage, we note the likeness of the physical objects called other people's bodies to each other and to our own body; we also note the likeness of their behaviour to our behaviour. In the case of our own behaviour, we can observe a number of correlations between stimulus and reaction (both being percepts). For example, we feel hunger or thirst, and then we eat or drink; we hear a loud noise, and we jump; we see Jones, and we say "Hullo, Jones." The behaviour of the percepts we call other people's bodies is similar to that of our own body in response to this or that stimulus; sometimes we experience the stimulus, and behave just as others do, which is the second stage; sometimes we do not experience the stimulus, but suppose, from their behaviour, that other people have experienced it, which is the third stage. This is a particularly plausible supposition if we ourselves experience the stimulus in question very shortly after we have observed the behaviour which led us to infer it. The third stage is the more important, since in the second we *might* attribute the behaviour of others to the stimulus which we perceive, and thus escape inferring unperceived existents, while in the third stage this alternative is not open to us. It will be seen that, in the third stage, the argument is the usual causal-inductive type of argument upon which all empirical laws are based. We perceive *A* and *B* conjoined in a number of cases, and we then infer *A* and *B* in a case in which we do not know by perception whether *A* is present or not. Moreover, the argument for other people's perceptions is the same in form and cogency as the argument for the future truth of laws of correlation among our own percepts. We have exactly as good reason for believing that others perceive what we do not as we have for believing that we shall have a perception of touch if we stretch out our hand to an object which looks as if it were within reach.

The argument is not demonstrative, either in the one case or in the other. A conjuror might make a waxwork man with a gramophone inside, and arrange a series of little mishaps of which the gramophone would give the audience warning. In dreams, people give evidence of being alive which is similar in kind to that which they give when we are awake; yet the people we see in dreams are supposed to have no external existence. Descartes' malicious demon is a logical possibility. For these reasons, we may be mistaken in any given instance. But it seems highly improbable that we are *always* mistaken. From the observed correlation of *A* and *B* we may argue, as regards cases in which *B* is observed but we do not know whether *A* exists or not, either: (1) *A* is always present, or (2) *A* is generally present, or (3) *A* is sometimes present. Dreams suffice to show that we cannot assert (1). But dreams could be distinguished from waking life by a solipsist, unless his dreams were unusually rational and coherent. We may therefore exclude them before beginning our induction. Even then, it would be very rash to assert (1). But (2) is more probable, and (3) seems extremely probable. Now (3) is enough to allow us to infer a proposition of great philosophic importance, namely: there are existents which I do not perceive. This proposition, therefore, if induction is valid at all, may be taken as reasonably certain. And, if so, it increases the probability of other propositions which infer the existence of this or that unperceived existent. The argument, though not demonstrative, is as good as any of the fundamental inductions of science.

We have been considering hitherto, not the external world in general, but the percepts of other people. We might say that we have been trying to prove that other people are alive, and not mere phantoms like the people in dreams. The exact thing we have been trying to prove is this: Given an observed correlation among our own percepts, in which the second term is what one would naturally call a percept of our own bodily behaviour, and given a percept of similar behaviour in a physical object not our own body but similar to it, we infer that this behaviour was preceded by an event analogous to the earlier term in the observed correlation among our percepts. This inference assumes nothing as to the distinction of mind and body or as to the nature of either.

In virtue of the above argument, I shall now assume that we may enlarge our own experience by testimony—*i.e.* that the noises we hear when it seems to us that other people are talking do in fact express something analogous to what we should be expressing if we made similar noises. This is a particular case of the principle contained in the preceding paragraph. I think the evidence for other people's percepts is the strongest we have for anything that we do not perceive ourselves; therefore it seems right to establish this, so far as we can, before proceeding to consider our evidence for "matter" —*i.e.* for existents satisfying the equations of physics. This must be our next task; but it will be well to begin with common-sense material "things" conceived as the causes of perceptions.

Having now admitted the percepts of other people, we can greatly enlarge the group constituting one "physical object." Within the solipsistic world, we found means of collecting groups of percepts and calling the group one physical object; but we can now enrich our group enormously. A number of people sitting near each other can all draw what they see, and can compare the resulting pictures; there will be similarities and differences. A number of stenographers listening to a lecture can all take notes of it, and compare results. A number of people can be brought successively into a room full of hidden roses, and asked "What do you smell?" In this way it appears that the world of each person is partly private and partly common. In the part which is common, there is found to be not identity, but only a greater or less degree of similarity, between the percepts of different people. It is the absence of identity which makes us reject the naive realism of common sense; it is the similarity which makes us accept the theory of a common origin for similar simultaneous perceptions.

The argument here is, I think, not so good as the argument for other people's percepts. In that case, we were inferring something very similar to what we know in our own experience, whereas in this case we are inferring something which can never be experienced, and of whose nature we can know no more than the inference warrants. Nevertheless, the common-sense arguments for an external cause of perception are strong.

To begin with, we can, without assuming anything that no one perceives, establish a common space and time in which we all live. (Our discussion is necessarily confined to people on the surface of the earth, since other people, if they exist, have not succeeded in communicating with us; consequently the complications of relativity do not yet arise.) The usual methods of determining latitude and longitude can be applied, without assuming that the readings of clock and sextant have the physical meaning usually assigned to them. Altitudes, also, can be measured by the usual methods. By these means, observers can be arranged in a three-dimensional order. Of course the resulting space will not be a continuum, since it will contain only so many "points" as there are observers. But the motion of an observer can be sensibly continuous, so that we can construct "ideal" points of view with defined mathematical properties, and thus build up, for mathematical purposes, a continuous space. We can thus arrive at the laws of perspective, taken in a generalized sense; that is to say, we can correlate the differences between correlated perceptions with differences in the situations of the percipients. And in the space derived from "points of view" we can place physical objects. For, let *A* and *B* be two observers, *a* and *b* their correlated visual percepts, which, being correlated, are described as percepts of one physical object *O*. If the angular dimensions of *a* are larger than those of *b*, we shall say (as a definition) that *A* is nearer to *O* than *B* is. We can thus construct a number of routes converging on *O*. We shall construct our geometry so that they intersect, and shall define their intersection as the place where *O* is. If *O* happens to be a human body, we shall find that the place of *O*, so defined, is identical with the place of *O* as an observer in the space of points of view.*

The correlation of the times of different percipients offers no difficulty, since, as before observed, our percipients are all on the earth. The usual method of light-signals can be employed. But here we come upon one of the arguments for the causal theory of perception, as against both common sense

* On this subject, *cf.* my *Knowledge of the External World*.

and phenomenalism. (We may define phenomenalism, at least for the moment, as the view that there are only percepts.) Suppose a gun on a hilltop is fired every day at twelve o'clock: many people both see and hear it fired, but the further they are from it the longer is the interval between seeing and hearing. This makes it very difficult to accept a naively realistic view as to the hearing, since, if that view were correct, there would have to be a fixed interval of time (presumably zero) between the sight and the sound. It also makes it natural to adopt a causal view of sound, since the retardation of the sound depends upon the distance, not upon the number of intermediate percipients. But hitherto our space was purely "ideal" except where there were percipients; it seems odd, therefore, that it should have an actual influence. It is much more natural to suppose that the sound travels over the intervening space, in which case something must be happening even in places where there is no one with ears to hear. The argument is perhaps not very strong, but we cannot deny that it has *some* force.

Much stronger arguments, however, are derivable from other sources. Suppose a room arranged with a man concealed behind a curtain, and also a camera and a dictaphone. Suppose two men came into the room, converse, dine, and smoke. If the record of the dictaphone and the camera agrees with that of the man behind the curtain, it is impossible to resist the conclusion that something happened where they were which bore an intimate relation to what the hidden man perceived. For that matter, one might have two cameras and two dictaphones, and compare their records. Such correspondences, which are only more extreme forms of those with which primitive common sense is familiar, make it inconceivably complicated and unplausible to suppose that nothing happens where there is no percipient. If the dictaphone and the hidden man give the same report of the conversation, one must suppose some causal connection, since otherwise the coincidence is in the highest degree improbable. But the causal connection is found to depend upon the position of the dictaphone at the time of the conversation, not upon the person who hears its record. This seems very strange, if its record does not exist until it is heard, as we shall have to suppose if we confine the world to percepts. I will not emphasize the more obvious oddities of such a world, as, *e.g.*, the one once brought forward by Dr G. E. Moore, that a railway train would only have wheels when it is not going, since, while it is going, the passengers cannot see them.

Before accepting such arguments, however, we must see what could be said against them by a phenomenalist. Let us, therefore, proceed to state the case for phenomenalism.

It may be suggested that our argument is, after all, not so strong as it looks, since all the facts can be interpreted by means of " ideal " percipients. The doubt I have in mind is suggested by a certain kind of construction, of which a good example is the introduction of " ideal " points, lines, and planes in descriptive geometry.* For our purposes, " ideal " points will suffice. The process by which they are constructed is as follows. Take all the straight lines which pass through a given point; these form a group of lines having other notable properties besides that of all possessing a common point. These other properties belong also to certain groups of lines which have no point in common—*e.g.* in Euclidean geometry, to the group consisting of all lines parallel to a given line. We then define a group of lines possessing these properties as an " ideal " point.† Thus some " ideal " points correspond to

* See Dr Whitehead's tract on this subject (Cambridge University Press). Also Pasch, *Neuere Geometrie*, Leipzig, 1882.

† The definition of an " ideal " point is as follows. Let l, m be any two lines in one plane, A any point not in this plane. Then the planes Al, Am have a line in common, say n. The class of all such lines as n, when A is varied while l and m remain fixed, is the " ideal " point determined by the two lines l, m.

real points, while others do not. In this way, by proceeding to "ideal" lines and planes, we arrive at last at a projective geometry, in which any two planes have a common line, and any two lines in a plane a common point, which immensely simplifies the statement of our propositions.

The analogy with our problem is perhaps closer than might be thought. We have, in the first place, real percepts, collected into groups each of which is defined by the characteristic that common sense would call all its members percepts of one physical object. These real percepts, as we saw, vary from one percipient to another in such a way as to allow us to construct a space of percipients, and to locate physical objects in this space. Let us, for the moment, adopt the view that nothing exists except percepts, our own and other people's. We shall then observe that the percepts forming a given group can always be arranged about a centre in the space of percipients, and we can fill out the group by interpolating "ideal" percepts, continuous in quality with actual percepts, in regions where there are no actual percipients. (A region of space which is "ideal" at one moment may be actual at another owing to motion of a percipient. The successive positions of an observer watching Cleopatra's Needle from a passing tram form a sensibly continuous series.) If a number of people hear a gun fired, there are differences in the loudness and the time of their percepts; we can fill out the actual percepts by "ideal" noises varying continuously from one actual one to another. The same can be done with correlated visual percepts; also with smells. We will call a group thus extended by interpolation and extrapolation a "full" group: its members are partly real, partly ideal. Each group has a centre in the space of percipients; this centre is real if occupied by a percipient, while otherwise it is ideal. (Our space is not assumed to be a smooth geometrical space, and the centre may be a finite volume.) As a rule, even when the centre is occupied by a percipient, it nevertheless contains no member of the group, not even an ideal member: " the eye sees not itself." A group, that is to say, is hollow: when we get sufficiently near to its centre it ceases to have members. This is a purely empirical observation.

A full group which contains any real members will be called a " real " group; a group whose members are all ideal will be called " ideal." It remains to show how we are to define an ideal group.

In addition to the laws correlating percepts forming one group—which may be called, in an extended sense, laws of perspective—there are also laws as to the manner in which percepts succeed one another. These are causal laws in the ordinary sense; they are included in the usual laws of physics. When we know a certain number of members of a full group, we can infer the others by the laws of perspective; it is found that some exist and some do not, but all that do exist are members of the calculated full group. In like manner, when we are given a sufficient number of full groups, we can calculate other full groups at other times. It is found that some of the calculated full groups are real, some ideal, but that all real groups are included among those calculated. (I am assuming an impossible perfection of physics.) Two groups belonging to different times may, in virtue of causal relations which we shall explain when we come to discuss substance, be connected in the way which makes us regard them as successive states of one " thing " or " body." (The time of a full group, by the way, is not exactly the time at which its members occur, but slightly earlier than the earliest real member—or much earlier, in the case of a star. The time of a full group is the time at which physics places the occurrence supposed to be perceived.) The whole series of groups belonging to a given " thing " is called a " biography." The causal laws are such as to allow us sometimes to infer " things." A thing is " real " when its biography contains at least one group which is "real," *i.e.* contains at least one percept; otherwise a thing is "ideal." This construction is closely analogous to that of "ideal" points, lines, and planes in descriptive geometry. We have to ask ourselves whether there are any reasons for or against it.

The above construction preserves the whole of physics, at least formally; and it gives an interpretation, in terms of percepts and their laws, to every proposition of physics which there is any empirical reason to believe. "Ideal" percepts, groups, and things, in this theory, are really a shorthand for stating the laws of actual percepts, and all empirical evidence has to do with actual percepts. The above account, therefore, preserves the truth of physics with the bare minimum of hypothesis. Of course there should be also rules for determining when a calculated percept is real and when it is ideal; but this is difficult, since such rules would have to contain a science of human actions. It may be known that you will see certain things if you look through a telescope, but it is difficult to know whether you will look through it. This completion of our science is therefore not possible at present; but that is no argument against the truth of our science so far as it goes. It is obvious that the method might be extended so as to make all perceptions except one's own "ideal"; we should then have a completely solipsistic interpretation of physics. I shall, however, ignore this extension, and consider only that form of the theory in which all percepts are admitted.

The metaphysic which we have been developing is essentially Berkeley's: whatever is, is perceived. But our reasons are somewhat different from his. We do not suggest that there is any impossibility about unperceived existents, but only that no strong ground exists for believing in them. Berkeley believed that the grounds against them were conclusive; we only suggest that the grounds in their favour are inconclusive. I am not asserting this: I am proposing it as a view to be considered.

The great difficulty in the above theory of "ideal" elements is that it is hard to see how anything merely imaginary can be essential to the statement of a causal law. We have to explain the dictaphone which repeats the conversation. We will suppose that it was seen in place before and after the conversation, but not during it. Consequently, on the view we are examining, it did not exist at all during the conversation. Causal laws, stated without fictitious elements, will thus involve action at a distance in time and space. Moreover, our percepts are not sufficient to determine the course of nature: we derive causal laws from close observation, and preserve them in other cases by inventing "ideal" things. This would not be necessary if percepts sufficed for the causal determination of future percepts. Thus the view we are examining is incompatible with physical determinism, in fact though not in form. We could multiply difficulties of this kind indefinitely. No one of them is conclusive, but in the aggregate they suffice to account for the fact that it is almost impossible to compel oneself to believe such a theory. Perhaps continuity (not in a strict mathematical sense) is one of the strongest objections. We experience sensible continuity when we move our own body, and when we fixedly observe some object which does not explode. But if we repeatedly open and shut our eyes we experience visual discontinuity, which we find it impossible to attribute to the physical objects which we alternately see and do not see, the more so as, to another spectator, they remain unchanged all the time. Causation at a distance in time, though not logically impossible, is also repugnant to our notions of the physical world. Therefore, although it is logically possible to interpret the physical world in terms of ideal elements, I conclude that this interpretation is unplausible, and that it has no positive grounds in its favour.

Nevertheless the above construction remains valid and important, as a method of separating perceptual and non-perceptual elements of physics, and of showing how much can be achieved by the former alone. As such, I shall continue to utilize it in the sequel. The only thing rejected is the view that "ideal" elements are unreal.*

The matter would, of course, be otherwise in this last respect if we could accept the argument for idealism, whether of the Berkeleyan or the German variety. These arguments profess to prove that what exists *must* have a mental character, and therefore compel us to interpret physics accordingly. I reject such *a priori* argumentation, whatever conclusion it may be designed to prove. There is no difficulty in interpreting physics idealistically, but there is also, I should say, no necessity for such an interpretation. "Matter," I shall contend, is known only as regards certain very abstract characteristics, which might quite well belong to a manifold of mental events, but might also belong to a different manifold. In fact, the only manifolds known for certain to possess the mathematical properties of the physical world are built up out of numbers, and belong to pure mathematics. Our reason for not regarding "matter" as actually being an arithmetical structure derived from the finite integers is the connection of "matter" with perception; that is why our present discussion is necessary. But this connection, as I shall try to show, tells us extremely little about the character of the unperceived events in the physical world. Unlike idealists and materialists, I do not believe that there is any other source of knowledge from which this meagre result can be supplemented. Like other people, I allow myself to speculate; but that is an exercise of imagination, not a process of demonstrative reasoning.

I shall assume henceforth not only that there are percepts

* The character of the "ideal" elements, also, will be less similar to that of percepts than in the above construction, or at least cannot be known to be so similar.

which I do not perceive, connected with other people's bodies, but also that there are events causally connected with percepts, as to which we do not know whether they are perceived or not. I shall assume, *e.g.*, that if I am alone in a room and I shut my eyes, the objects in it which I no longer see (*i.e.* the causes of my visual percepts) continue to exist, and do not suddenly become resurrected when I re-open my eyes. This must be taken in conjunction with what was said earlier about perspective in a generalized sense, and about the common space in which we locate the physical objects which, for common sense, are perceived by several people at once. We collect correlated percepts into a group, and we suppose that there are other members of the group, corresponding to places where there is no percipient—or, to speak more guardedly, where there is not known to be a percipient. But we no longer assume, as when we were constructing " ideal " elements, that what is at such places is what we should perceive if we went to them. We think, *e.g.*, that light consists of waves of a certain kind, but becomes transformed, on contact with the eye, into a different physical process. Therefore what occurs before the light reaches an eye is presumably different from what occurs afterwards, and therefore different from a visual percept. But it is supposed to be causally continuous with the visual percept; and it is largely for the sake of this causal continuity that a certain reinterpretation of the physical world seems desirable.

In some ways, the language of causation is perhaps not the best for expressing what is intended. What is intended may be expressed as follows. Confining ourselves, to begin with, to the percepts of various observers, we can form groups of percepts connected approximately, though not exactly, by laws which may be called laws of " perspective." By means of these laws, together with the changes in our other percepts which are connected with the perception of bodily movement, we can form the conception of a space in which percipients are situated, and we find that in this space all the percepts belonging to one group (*i.e.* of the same physical object, from the standpoint of common sense) can be ordered about a centre, which we take to be the place where the physical object in question is. (For us, this is a *definition* of the place of a physical object.) The centre is not to be conceived as a point, but as a volume, which may be as small as an electron or as large as a star. The essential assumption for what is commonly called the causal theory is, that the group of percepts can be enlarged by the addition of other events, ranged in the same space about the same centre, and connected both with each other and with the group of percepts by laws which include the laws of perspective. The essential points are (1) the arrangement about a centre, (2) the continuity between percepts and correlated events in other parts of the space derived from percepts and locomotion. The first is a matter of observation; the second is a hypothesis designed to secure simplicity and continuity in the laws of correlation suggested by the grouping of percepts. It cannot be demonstrated, but its merits are of the same kind as those of any other scientific theory, and I shall therefore henceforth assume it.

CHAPTER XXI

PERCEPTION AND OBJECTIVITY

WHEN a number of people are, from the standpoint of common sense, observing the same object, there are both likenesses and differences among their percepts. For common sense, with its naive realism, the differences constitute a difficulty, since they render the percepts mutually inconsistent if taken to be each wholly a revelation of one and the same physical object. But to the causal theory of perception this difficulty is non-existent. We have now, however, an opposite difficulty—namely, that of deciding what elements in a percept can be used for inference as to the existence of something other than itself, and as to the nature of the inferences when they can be drawn. For the moment, I am not thinking of inferences involving motion, but only of inferences as to the present state of the physical object which is being observed.

We must be on our guard against a confusion which is difficult to avoid in such inquiries. Perception, as an event in our own history, is a recognizable occurrence; its psychoogical meaning is fairly definite. But it has also an epistemological meaning, and this is hardly capable of being made as definite as could be wished. Perception is interesting to us, in our present discussion, because it is a source of knowledge, not because it is an occurrence which a psychologist can recognize. So long as naive realism remained tenable, perception was knowledge of a physical object, obtained through the senses, not by inference. But in accepting the causal theory of perception we have committed ourselves to the view that perception gives no immediate knowledge of a physical object, but at best a datum for inference. A perception does, however, still give knowledge of something: if I perceive a round red patch, I know that there is a round red patch in the world now, and no account of the causes of my perception can destroy this knowledge. It may be conceded that, in saying this, I am using "perception" more narrowly than it might be used in psychology: I am confining it to cases where we notice explicitly what we are perceiving. For epistemological purposes, this restriction is essential. I am deliberately refraining from all analysis of "knowing," since that would take us too far from our subject.

The inferences to be primarily drawn from a perception are as to other members of the group to which the percept concerned belongs. This is done, in a confused way, by common sense, when it infers the "real" size or shape of an object from its "apparent" size or shape, *i.e.* from the real size or shape of the percept. The "real" size or shape is a norm, from which the percept of a spectator in a given relative situation can be inferred. Ordinarily, there is no conscious inference involved; but conscious inference can be used without invoking any fresh knowledge. For example, an architect can show the view of a proposed house from any angle when he knows its measurements, and for this purpose he uses only systematized common sense; and he can infer the measurements approximately when he has viewed an actual house from several angles. The "real" object, as opposed to its "appearances," is thus something of the nature of a formula by means of which all sufficiently near "appearances" can be determined. Given the measurements of a house, we can infer its apparent shape at a given distance in a given direction. If perception were perfectly accurate and regular, a few percepts belonging to a given group would enable us to determine all percepts, actual and possible, belonging to that group.

This is found to be not in fact the case. From seeing a drop of water with the naked eye, we cannot know that under the microscope it will be found to be full of bacilli. When we see a man a hundred yards away, we cannot tell whether he is handsome or plain. When we can only just distinguish a person's voice, we cannot tell what is being said. These are all cases of "vagueness," in a certain perfectly precise sense. In any group of percepts, those nearer the centre have a many-one relation to those farther off—*i.e.* two things which look alike from a distance look different when seen close to. In this sense, the more distant percepts are vaguer than the nearer ones: the former can be inferred from the latter, but not the latter from the former.

There is, however, a converse fact—namely, that what may be called the "regular" law for inferring distant from near appearances may be interfered with by intervening things. The sun may be visible from a great altitude when clouds make it invisible from the earth's surface. Sounds may be stopped by obstacles, and die away completely at a sufficient distance from their source. Smells die away still more quickly, and are even more dependent upon the wind. This set of facts interferes with the inference from near to distant appearances, just as the former set interfered with the inference from distant to near appearances.

There is, however, an important difference between the two sets of facts. The increasing vagueness of distant appearances is an intrinsic law of groups of percepts, whereas the uncertainty as to distant appearances when near appearances are given depends always upon outside interference. This distinction is of a kind which we shall find to be very important in various ways. Let us try to state it clearly in the case in question.

Suppose two persons to be both observing a given object which is stationary on the earth's surface, and suppose that one of the persons remains at rest while the other moves about. We will suppose that to the person who remains at rest there is no perceptible change in the object throughout the time concerned. To the other person there will be changes which, in general, are approximately according to the laws of perspective, especially for small changes in the observer's position. But sometimes, to take the most obvious example, the object in question becomes invisible when the observer takes up certain positions—those, namely, from which some opaque object is between the observer and the object which he had been seeing. As a rule, this happens gradually: at first both objects are visible, gradually their angular distance becomes less, and at last only the nearer object remains visible. The nearer object has thus had an effect upon the appearance of the farther object. Fog, smoke, glass, blue spectacles, etc., similarly modify the appearances of distant objects. That is to say, in calculating the appearance which a body will present in such and such a place, we have to take account, not only of the body's appearances elsewhere, but also of the bodies between it and the place in question. These intervening bodies are sometimes sensible, sometimes not; when they are not, they are inferred as being necessary in order to preserve the laws which have been found to hold when they were sensible. The principle is the following: If we compare neighbouring members of a group of percepts, we find, in a great many cases, that their first-order differences are in accordance with the laws of perspective, while their second-order differences are functions of groups with other centres; or rather, since the above statement is too precise for the facts, we may say simply that the differences between neighbouring positions are compounded of the laws of perspective together with functions of groups with other centres. Suppose, *e.g.*, that you are seeing an object through glass which is slightly distorting. The glass is a tactual group between you and the object; as you move, the distortions due to the glass change, and have to be compounded with the laws of perspective in order to calculate one member of a group from another. In other cases, by carefully comparing a number of members of a group, we can discover that their departure from perspective laws proceeds according to a law which is a function of a position not perceptibly occupied. The previous illustration will apply to this case also, if we have not touched the distorting glass. Human beings are superior to birds and insects in the fact that they can infer glass in such cases, without any scientific apparatus, whereas birds and insects repeatedly bump into it.

Like much of what has to be said in the transition from perception to science, the above statement is not capable of being made in an exact form. The methods by which we collect a number of percepts into one group are rough and ready, and become impossible if there is very great distortion by the intervening medium. But these methods are successful in a sufficient number of cases to give rise to the notion of events grouped about a centre, changing partly in accordance with the laws of perspective and partly in ways which are functions of groups with other centres. Having arrived at this notion, it is not very difficult to modify it in such a way that it shall become capable of scientific precision.

I come now to the question of "objectivity" in a perception. This is a matter of degree: the more correct are the inferences we can draw from a percept as to other events (whether percepts or not) belonging to the same group, the more "objective" is the perception. (I propose this as a definition.) A percept may not belong to a group at all; in that case it has no objectivity. Hallucinations and dreams come under this head. Or we may be mistaken as to the position of the centre of the group; this is the case with a mirage, or with a reflection not recognized as such. Or we may perceive a colour or shape which is erratic, say owing to intervening smoke, and thus misleads us as to the colour or shape which others will see. I should not regard a perception as failing in objectivity through mere vagueness. Vagueness diminishes the number of inferences that we can draw, but not their correctness. From a distance we perceive correctly that what is approaching is a man; when he gets near we perceive that he is Jones. But our previous perception did not fail in objectivity through failing to show that it was Jones. It would have failed of objectivity if, owing to intervening lenses, it had shown us a man standing on his head.

When two people simultaneously have percepts which they regard as belonging to one group, if the inferences of the one differ from those of the other, one of them at least must be drawing false inferences, and must therefore have an element of subjectivity in his perception. It is only where the inferences of the two observers agree that both perceptions may be objective. It will be seen that, according to this view, the objectivity of a perception does not depend only upon what it is in itself, but also upon the experience of the percipient. A man accustomed to being short-sighted can judge objects much more correctly than a man whose vision suddenly acquires the same defect. Fatigue as well as alcohol may make us see double, but fatigue will not deceive us when it does so.

Subjectivity in perceptions may be traced to three sources, physical, physiological, and psychological; or, better perhaps, physical, sensory, and cerebral. In all cases in which a percept is really a member of a group constituting a physical object, any element of subjectivity that it may possess is due to the distortions connected with intervening physical objects —that, at least, is the theory which has been found successful. When these objects are between the body of the percipient and the centre of the group to which the percept belongs, the subjectivity is physical; when they are in the body of the percipient but not in his brain, they are sensory; when they are in his brain, they are cerebral. The last of these, however, is usually purely hypothetical; the *discoverable* causes of the subjectivity which we are calling cerebral are as a rule psychological.

Physical subjectivity exists equally in a photograph or gramophone record; it is present already in the events, external to the percipient's body, which belong to the group in question and are very near to the sense-organ concerned in the perception. The stick that looks bent when it is half in water is an obvious example of physical subjectivity. So are many effects of reflexion, refraction, etc. The theory of relativity has brought to light a new kind of physical subjectivity, dependent upon relative motion. The prevention of mistaken inferences owing to physical subjectivity is part of the business of physics, and does not involve physiology or psychology.

Physiological (or sensory) subjectivity arises through defects of the sense-organs or afferent nerves; it may also be produced by drugs. We can discover such defects by the comparison of different people's perceptions in a given situation. It should be observed that the intrinsic quality of a percept is unimportant in this respect: if one person sees red where another sees green, and green where another sees red, the fact will be undiscoverable and harmless. But if, where one person sees two colours, red and green, another only sees one, we have a discoverable difference, which is correctly described as a defect in the vision of the person who only sees one. It is always assumed that if two stimuli produce noticeably different effects in a given percipient at a given time, there must be differences in the stimuli correlated with the differences in their effects; while if the effects are not noticeably different, there may nevertheless be differences in the stimuli. Consequently *A*'s senses are better than *B*'s if *A* perceives differences when *B* does not. For the same reason, the microscope and the telescope are better than the naked eye. But this has, as a rule, more to do with vagueness than with subjectivity. Subjectivity only enters in when we are led to make false inferences, not when we are merely unable to make inferences which another can make. A mere deficiency, such as blindness or deafness, does not amount to subjectivity, but seeing double does if it deceives us. It deceives us when it leads to false inferences—*e.g.* that there are two tactual objects, or that a person near us will see two objects.

Cerebral (or psychological) subjectivity arises as a result of past experience. An obvious example is a sensation which appears to be in a leg which has been amputated. We are liable to this kind of error whenever two things usually associated are for some reason dissociated. Certain sensations have, in the past, been generally associated with a stimulus in the leg; but they have had as intermediaries conditions of the nerves between the leg and the brain. If these previously intermediate conditions arise in a person who has lost his leg, he will interpret them as sensations in his leg, if he has momentarily forgotten that he has lost his leg—*e.g.* on waking from sleep. In all perception (except perhaps during the first weeks of life) there is a large element of interpretation due to past experience, and this element is subjective when the present situation does not contain the correlations whose past occurrence has caused the interpretation.

All these sources of error have to be guarded against if perception is not to mislead us. The ways of guarding against them are those suggested by common sense and perfected by science; they are all such as to substitute laws with few or no exceptions for laws with a comparatively large number of exceptions.

It will be seen that very little can be inferred with confidence from a single percept; we need observation from different points of view, and throughout a certain period of time. It is true that we shall *usually* be right in what we infer from a single percept, but that is because the objects that surround us mostly belong to familiar kinds—men, horses, motor-cars, etc. But it would not be difficult to construct situations which would deceive at the first glance, especially if we could be suddenly transported into a quite unfamiliar world, like Wells's Martians. Water, for example, would completely puzzle a person who had never seen a liquid, if such a person could exist. In this matter, as elsewhere, we proceed step by step from the easy but precarious inferences of common sense to the difficult but more reliable inferences of science.

Where the intervening medium is relevant in inferring other members of a group from a percept, it is obvious that the single percept is theoretically inadequate as a basis for inference, since, by a change in the medium, the same percept might be associated with a different group. In this case, the distorting element in the medium may be directly discovered by other percepts—*e.g.* glass may be touched—or it may be merely inferred by examining the way in which percepts belonging to one group change from place to place—*e.g.* refraction in air. When it has been inferred, the inference needs to be tested by examining whether it has further consequences which can be verified. All this is a commonplace.

It remains to say something about the inference from percepts to events which no one perceives. It is not its validity that I wish to examine now, but its scope—*i.e.* how much we can know about unperceived events, assuming the causal theory of perception. It is sometimes urged that an unperceived cause of a perception must be a mere *Ding-an-sich* or Spencerian Unknowable. This seems to me only very partially true, if we accept the usual canons of scientific inference. We assume that differences in percepts imply differences in stimuli—*i.e.* if a person hears two sounds at once, or sees two colours at once, two physically different stimuli have reached his ear or his eye. This principle, together with spatio-temporal continuity, suffices to give a great deal of knowledge as to the *structure* of stimuli. Their intrinsic characters, it is true, must remain unknown; but we may assume that the stimuli causing us to hear notes of different pitches form a series in respect of some character which corresponds causally with pitch, and we may make similar assumptions in regard to colour or any other character of sensations which is capable of serial arrangement. And we can without difficulty extend geometry to the world outside our perceptions, although the space of that world will only correspond to the space of perception in certain respects, and will be by no means identical with the space of perception.

What we assume is, formally, something like this: there is a roughly one-one relation between stimulus and percept—*i.e.* between the events just outside the sense-organ and the event which we call a perception. This enables us to infer certain mathematical properties of the stimulus when we know the percept, and conversely enables us to infer the percept when we know these mathematical properties of the stimulus. Consequently, except when we are studying physiology or psychology, we may suppose that what is happening in a place is what a person would perceive in that place, provided we use, in inference, only those properties of the percept which it shares with the stimulus. *E.g.* we must not use the blueness of blue, but we may use its difference from red or yellow. We cannot argue that because a picture looks beautiful, therefore there is beauty in the system of stimuli, because beauty may depend upon the actual qualities.* But nothing in physical science ever depends upon the actual qualities. Hence for practical purposes in physics the difference between percept and stimulus only compels us to confine ourselves to the structural properties of percepts; so long as we do this,

* If we accepted the theory that beauty depends only upon "significant form," we should have to say that a musical score is as beautiful as the music which it represents.

we need hardly trouble to remember that percept and stimulus are different. In physiology and psychology this does not hold, since we are concerned with the process intervening between stimulus and perception, or with perception itself.

Even in physics, it does not hold strictly, because the relation of stimulus and perception is not strictly one-one. It is only approximately so, even when we confine ourselves to stimuli to a given sense of a given person at a given time—*e.g.* two colours which I perceive side by side. Even here, vagueness comes in, so that slightly different stimuli may give indistinguishable perceptions. This constitutes an essential limitation to our knowledge, enshrined in the notion of " probable error." It can, however, be reduced to a minimum by the usual methods and constitutes, therefore, rather a practical difficulty than a theoretical problem.

CHAPTER XXII

THE BELIEF IN GENERAL LAWS

THROUGHOUT our discussion of perception and the physical object, we have assumed the validity of general laws. This is always assumed in scientific practice, but the reasons for assuming it are not very clear. Although the subject is not one on which it is easy to say anything definite, yet it seems necessary to examine it.

Like other scientific postulates, the belief in general laws is rooted in the properties of nervous tissue—the same properties which make us believe in induction and enable us to learn from experience. This origin, of course, affords no warrant for the truth of the belief, but equally gives no reason against it. Indeed, so far as it goes, it affords a slight presumption in favour of the view that a great many events are in accordance with general laws, since it shows that animals which act in a way which the truth of this belief would render rational can survive. I should not wish, however, to lay stress upon such an argument.

When we first begin to think, we find ourselves acting in certain ways which seem to succeed, and we set to work to rationalize our behaviour. The natural way to do this is to say: Things *always* happen that way. This so often succeeds that we acquire the habit of always supposing that there is some general law according to which any particular event has occurred. This belief has two practical consequences. First, when a set of events are all in accordance with some law, we expect other similar events to be in accordance with it. Secondly, when a set of events appears irregular, we invent hypotheses to regularize it. Both procedures are important.

The first of these procedures is simply induction. As such, it is fundamental, in some form or other, and I propose to say no more about it.

The second is more interesting for our purposes. When an induction fails in a surprising way—*e.g.* when there is an eclipse—there are two things which a primitive man may do. He may regard the failure as a "portent," in no way invalidating the general validity of the induction, but showing that there is something strange, and probably terrifying, in the special circumstances connected with the astonishing event. Or he may look for some general law different from that which has hitherto proved adequate, in the hope that the new law may account for the exceptional occurrence as well. The latter course will seldom be adopted until a high degree of intellectual culture has been attained. If the odd event is on a large scale, it will be considered superstitiously, and if not, it will be simply ignored. Sometimes, however, a general law is found by accident, as a result of the careful records inspired by superstition. This evidently happened with the Egyptian priesthood, who learnt to predict eclipses, and probably only then ceased to regard them with awe. Gradually, the view that there must be *some* law according to which strange things happened became more widespread. Dr Whitehead, in his *Science and the Modern World*,* traces the belief in natural laws to various sources, such as: Fate in Greek tragedy, the supremacy of Roman law, and the rationality of God in mediæval theology. In effect, however, he regards the belief as having only acquired a firm hold of the scientific mind at the renaissance. Everything that he says on this subject is so excellent that it is unnecessary to cover the ground again.

Although the belief in the universality of natural law was, at the time of the renaissance, a bold faith going far in advance of the evidence, it has since been so successful that it is now

* Chap. i., especially p. 5 ff.

possible to defend it on inductive grounds. But there is some difficulty in deciding what we are to mean by it. I have dealt with this subject before,* and shall now consider it only briefly.

The regularities which we first observe, and in which we first believe, are of the simple form: "*A* is always accompanied (or preceded or succeeded) by *B*." But all such regularities are capable of having exceptions, and science soon seeks laws of a different kind. We arrive in the end (possibly not at the very end) at differential equations. I think that these are of two kinds, those expressing persistence, and those expressing accelerations (in a generalized sense). The former are concealed, more or less, by the assumption of permanent substance; but this is a topic which I shall consider in the next chapter. The latter are the ordinary differential equations of the second order which occur throughout mathematical physics. But in addition to these, in order to produce observed macroscopic results, there must be statistical laws governing quantum changes and radio-active disruptions of atoms. I want to inquire whether we are saying anything significant in assuming that there are laws governing the course of the physical world, or whether *any* set of percepts must be amenable to law by a sufficiently liberal use of hypothesis.

It is by no means clear that the accepted laws of physics make certain imaginable series of percepts impossible; still less that the mere existence of laws would have this effect. Take, *e.g.*, continuity. Changes which appear sudden (*e.g.* explosions) can be resolved into a number of continuous though rapid changes: *per contra*, situations in which there appears to be no change (*e.g.* a steadily glowing gas) are resolved into a number of discontinuous changes. Thus we can neither infer the absence of physical continuity from the absence of

* *Cf.* "On the Notion of Cause," in *Mysticism and Logic.*

continuity in percepts, nor the presence of physical continuity from the presence of continuity in percepts. Again: if percepts change in unexpected ways, we infer unperceived matter; and by a sufficient amount of unperceived matter almost any series of percepts could be explained. Of course a particular law is strengthened when it enables us to predict percepts, but this belongs to the arguments in favour of such-and-such laws, not to the arguments in favour of laws in general. We can have evidence in favour of such-and-such a law without having evidence for laws in general. But here we must make some distinctions. Evidence in favour of a particular law is evidence that a certain class of phenomena are subject to a rule which we have succeeded in discovering. If so, they are sure to be also subject to other rules sensibly indistinguishable from the one for which we have evidence; but these will in general be more complicated than the rule which we adopt. Complication may be of two kinds: it may be in the formula, or in the amount of hypothetical matter needed to make the rule work. The great merit of Newtonian gravitation was that it was simple in both respects. But clearly any set of observations on planetary motions could have been fitted into the Newtonian formula by postulating a sufficient number of invisible bodies or a sufficient complication in the law of attraction. For any given set of observations, there would have been many such possible methods of bringing harmony between observation and theory; most of these would not have been compatible with a fresh set of observations, but some of them would have been, given sufficient mathematical ingenuity. What is remarkable, therefore, is not the reign of law, but the reign of *simple* laws. If the transfer of energy were subject to laws as complicated as those governing the transfer of English land, we should never succeed in discovering them: there would always remain a number of possible codes, all of which would fit all known relevant facts.

The principle of induction, as practically employed, is the principle that the *simplest* law which fits the known facts will also fit the facts to be discovered hereafter. This principle, in all its naked simplicity, has come to the fore in Einstein's theory of gravitation, which consists in taking the simplest available tensor equation in preference to the others that are mathematically possible.

It may be said that the principle of simple laws is purely heuristic, and of course this is true to a considerable extent. No sensible mathematician would test a complicated formula before testing a simple one. But the remarkable thing is that the simple formula so often turns out right. From the trend of physics, it seems as though complication were geographical rather than legal. Organic compounds have an immensely complicated structure, but there is no reason to suppose that their fundamental laws are other than those which govern the hydrogen atom. Professor J. B. Haldane, it is true, thinks otherwise, and so do all varieties of vitalists. But, to a layman, their arguments seem inconclusive, and they are rejected by many competent authorities. It is therefore at least a tenable hypothesis that all matter is governed by very simple laws. This is so remarkable that it almost suggests some relation to Mr Keynes's "principle of limitation of variety," and seems to confirm his hint that Nature may be really like the urn containing white and black balls which plays such a prominent part in the theory of probability. Some Mendelians would make us think of human beings in this way. Suppose there were a hundred pairs of characters, a, a', b, b', c, c', etc., such that every human being possessed by inheritance one but not both of the characters in each pair. This would make the number of differing human embryos 2^{100} —*i.e.* about 10^{30}. If this is thought too few, we can take more pairs of characters. Views of this sort cannot be rejected out of hand, and they are strongly suggested by the success of induction and the prevalence of simple laws. Let us, therefore, ask once more: What evidence is there that simple laws prevail, and how much reason have we to be surprised by the degree of their prevalence ?

As I have pointed out on a former occasion, it would be fallacious to argue inductively from the simplicity of the laws we have discovered to the probable simplicity of undiscovered laws. For, if some laws are simple and some complicated, we are likely to discover the simple laws first. We have to proceed more cautiously. First, is it surprising that there are *any* simple laws ? Secondly, have we any ground for believing, as was suggested just now, that *all* phenomena are governed by simple laws ?

Simplicity is best established at the two opposite extremes of size: astronomy and the atom. The latter, however, is much more significant for our inquiry, since the simplicity of astronomy may result from averaging. As we saw in Part I., the theory of the atom amounts, broadly, to this: An atom is composed of electrons and protons, the latter being all in the nucleus, the former partly in the nucleus (except in hydrogen), partly planetary. The number of protons in the nucleus gives the atomic weight; the excess of the number of protons over that of electrons in the nucleus gives the atomic number. When the atom is unelectrified, the number of planetary electrons is equal to the atomic number. If the quantum theory is correct, an atom has a certain number of characters, each measured by integers called quantum numbers, which are always small. It has also a property called energy, which is a function of the quantum numbers; and in connection with each of the quantum numbers there is a periodic process which is subject to quantum rules. Each quantum number is capable of changing suddenly from one integer to another. When the atom is left to itself, these changes will only be such as to diminish the energy, but when it is receiving energy from elsewhere the changes may increase the energy. All this, however, is more or less hypothetical. What we really know about is the interchange of energy between the atom and the surrounding space; here there are simple laws as to the form the radiant energy will take. But there are at present no laws determining *when* quantum changes will take place in the atom, though the changes that are possible are a definite known set.

As we are only considering how far simple laws *can* account for the phenomena, we may accept the view of the atom as a miniature solar system, governed, except as to quantum changes, by attractions and repulsions among its electrons. Nevertheless it remains a fact that the atom only indicates its presence when it suffers a quantum change, and that we know of no laws determining why, at a given moment, such a change takes place in some atoms rather than in others. The laws governing the intensity of the light emitted by a gas are statistical laws. This suggests a world in which the number of possibilities is finite, but the choice among possibilities is left purely to chance. We might suppose, as Poincaré once suggested, and as Pythagoras apparently believed, that space and time are granular, not continuous—*i.e.* the distance between two electrons may be always an integral multiple of some unit, and so may the time between two events in the history of one electron. This, together with the fact that the number of electrons is finite, would give a finite number of possible situations for each electron. And it may be that the choice among possible situations is wholly a matter of chance. In that case, the apparent regularity of the world will be due to the *absence* of laws. I think it improbable that such a view could be developed satisfactorily, but at least we must take account of it before we attach undue importance to the appearance of law in the world.

The real objection to a philosophy founded upon such a

theory of the universe as we have been considering is that, after all, we still need statistical laws, which will involve a "random distribution," or something of the kind. Such laws are still laws, though they differ from others by seeming *a priori* probable instead of improbable. To this extent, it is a gain if we can base science upon them; but it would not be correct to say that, in that case, science would have succeeded in doing without laws. We could no longer say, however, that the laws of science were surprising; on the contrary, we should be surprised by their failure.

There is another question to be considered, and that is as to the scope of simple laws. It cannot be pretended that we *know* the laws governing the hydrogen atom to be sufficient to account for all that happens to matter, especially to organic matter. This is at present merely a hypothesis. All science uses laws based upon observation, which may or may not be deducible by a celestial mathematician from the laws governing electrons, but are not likely ever to be deducible by mathematicians on this planet. And when we come to such matters as physiology, the laws are no longer such as to enable us to say, with any confidence, just what is going to happen; they give tendencies rather than precise mathematical rules. It would be rash to maintain that such rules must exist; we may do well to look for them, but not well to feel quite certain that they are to be found.

On the whole, the tendency of the foregoing discussion has been to suggest that it is easy to exaggerate the evidence for simple laws in the physical world. Where we know most—*i.e.* in regard to the structure of the atom—there is, so far as we know, a complete absence of law in certain very important respects. Where we know less, the laws may be purely statistical. The amount of law known to exist in the physical world is, therefore, less surprising than it seems at first sight, and there is no conclusive reason for believing that all natural occurrences happen in accordance with laws which suffice to determine them given a sufficient knowledge of their antecedents. Science must continue to postulate laws, since it is coextensive with the domain of natural law. But it need not assume that there are laws everywhere; it need only assume, what is evident since it is a tautology, that there are laws wherever there is science.

CHAPTER XXIII

SUBSTANCE

THE question of substance in the philosophy of physics has three branches: logical, physical, and epistemological. The first is a problem in pure philosophy: is the notion of " substance " in any sense a " category," *i.e.* forced upon us by the general nature either of facts or of knowledge ? The second is a question of the interpretation of mathematical physics: is it (*a*) necessary or (*b*) convenient to interpret our formulæ in terms of permanent entities with changing states and relations ? The third concerns the special topics with which we are concerned in Part II.—namely, the relation of perception to the physical world. The first and second problems really belong to other portions of the philosophy of matter, but I shall discuss them here in order to obtain a unified discussion of the problem of substance.

Logically, " substance " has played a very important part in the past, and is still perhaps less obsolete than might be supposed. A substance may be defined in purely logical terms as " that which can only enter into a proposition as subject, never as predicate or relation." This definition is practically that of Leibniz, except that he does not mention relations, since he held them to be unreal. We shall do well, however, to include them, because the logical position of substance is not much affected thereby, and it may, I hope, be now taken for granted that relations are as "real" as predicates.

Metaphysically, substances have generally been held to be indestructible. But this opinion is not justified by the logical definition, though many philosophers have supposed that it was. When I wish to discuss a substance having this further attribute, I shall speak of it as a "permanent substance"; when I use the word "substance" without qualification, I shall mean only substance in the logical sense, leaving the question of duration open.

It is extraordinarily difficult, in considering substance from the point of view of logic, to avoid being unduly influenced by the structure of language. All languages commonly known to civilized people consist of sentences which can be analyzed into subject and predicate, two subjects and a dyadic relation, three subjects and a triadic relation, etc., together with relations between such units, expressed by "or" or "if" or some analogous word. I do not know whether the same can be said of African, Australian, or other uncivilized languages. But certainly it can be said of all the languages that philosophers have known. Logic, as ordinarily conceived, takes over this linguistic scheme, and is inclined to attribute metaphysical importance to it. We can hardly resist the belief that the structure of the sentence reproduces the structure of the fact which it asserts, or, in the case of false sentences, of the fact which would exist if the assertion were true. This belief, natural as it is, seems very unplausible when explicitly stated. Nevertheless, I believe that it has some element of truth, though it is very hard to disentangle this element. An attempt was made by Wittgenstein,* and I have been much influenced by his point of view.

If we admit, as it seems natural to do, that some sentences, taken in their usual meaning, correspond to facts, while others do not, we must suppose that the structure of sentences is related, in some way, to the structure of facts, since otherwise such correspondence would be impossible. Moreover, a sentence is a physical fact, and may therefore be expected to be capable of correspondence with other physical facts. These

* *Tractatus Logico-Philosophicus.*

two arguments come from quite different intellectual regions, the one being logical, the other physical. If we were discussing anything other than physics, they would work in opposite directions, and tend to show that we cannot understand (at least verbally) anything having a structure radically different from that of events in space-time. For our purposes, however, the two arguments are concurrent.

Let us, for a moment, consider a sentence as a physical occurrence. We must distinguish between spoken and written sentences, since the former are evanescent events while the latter are pieces of matter. We must also distinguish between a sentence in the sense in which it is unique on each occasion when it is uttered or written, and a sentence in the sense in which the same sentence occurs at a given place in each copy of the same book. *E.g.* Jeremiah xvii. 9 is a sentence in the latter sense; in the former sense, the particular series of shapes at that point in my Bible constitute a sentence, while those in yours constitute another (similar) sentence. The former sense comes first when we are considering a sentence as a physical occurrence; the latter, when we are considering it as having "meaning."

A spoken sentence, considered physically, is a series of noises from the point of view of the hearer, and a series of movements in the mouth and throat from the point of view of the speaker. The "meaning" of the sentence depends upon the causes of the spoken words and the effects of the heard words.* But for the moment let us ignore "meaning." Then we find that the sentence consists essentially of noises in order: the order is as essential as the character of the noises. (In a language like Latin, this is not so true of the separate words as in a modern language, but it is just as true of the parts of words: "Roma" is a different word from "amor.") Considered as physical occurrences, the words expressing different

* Cf. *Analysis of Mind*, chap. x.

parts of speech are indistinguishable; nevertheless there are relations which are symbolized by relations among words, not by words. Consider " Brutus killed Cæsar " and " Cæsar killed Brutus." The difference between these two statements is indicated, in an unniflected language, not by a word, but by a relation among words. Thus a spoken sentence consists of certain noises in a certain temporal order. In the sentence, we can distinguish terms and relations: the terms are the words (or, more strictly, the elementary noises which, in a phonetic system, would each be represented by a separate letter), and the relations are temporal relations among events. According to our definition, the elementary noises composing the sentence may count as " substances," in spite of the fact that they are evanescent.

In the case of written words, the sentence is no longer a temporal series of events, but a spatial series of material structures. It is not essential to a written sentence that its parts should stand for sounds: in some languages (*e.g.* Chinese) this is not the case, and there is some reason to think that writing developed from pictures, not from the attempt to symbolize speech. We may therefore treat the written language as an independent method of conveying meaning. It is obvious that its efficacy in this respect depends upon its capacity for causing visual perceptions (or tactual perceptions in the case of " Braille "). Written words, even Chinese ideograms, consist essentially of parts with a structure, and the structure is essential to the meaning. This is equally the case with a sentence, even in Latin. Take " Cæsar amat Brutum " and " Cæsarem amat Brutus." Here the case-endings may be regarded as separate words (which they probably were originally), whose position relative to the stem " Brut " or " Cæsar " indicates the " sense " of the relation asserted.

The written language depends upon the causal theory of

perception and the existence of physical objects; the spoken language involves the former, but not the latter. Thus in the written language the "substantial" elements have a permanence (throughout some finite time) which they do not have in the spoken language. Their permanence, however, is not metaphysical or absolute; it is only like that of houses or trees. It depends upon the fact that matter arranged in certain patterns will often retain those patterns for a long time, though not for ever. And the essential thing about writing is its capacity for causing visual events.

So far, we have seen no reason to suppose that the suggestions of language are misleading where the physical world is concerned, since language is a physical phenomenon, and must share whatever structure all such phenomena have in common. But the philosophy which has been based on language—or, perhaps, has moulded language—has further elements which are more dubious. These are derived from the distinctions between parts of speech. Philosophers have, as a rule, failed to notice more than two types of sentence, exemplified by the two statements "this is yellow" and "buttercups are yellow." They mistakenly supposed that these two were one and the same type, and also that all propositions were of this type. The former error was exposed by Frege and Peano; the latter was found to make the explanation of order impossible. Consequently the traditional view that all propositions ascribe a predicate to a subject collapsed, and with it the metaphysical systems which were based upon it, consciously or unconsciously. This did away with the objections to pluralism as a metaphysic.

But there remain certain linguistic distinctions which *may* have metaphysical importance. There are proper names, adjectives, verbs, prepositions, and conjunctions. It is natural to hold that, in an ideal language, proper names would indicate substances, adjectives would indicate the properties

by means of which substances are collected into classes, verbs and prepositions would indicate relations, and conjunctions would indicate the relations between propositions by means of which we build up what are called "truth-functions."*
If there really are these categories in the world, it is desirable that language should symbolize them, and metaphysical errors are likely to result if language performs this task inaccurately. For my part, I believe that there are such categories, except, perhaps, conjunctions. But I will not argue the question at this point, since I wish, as far as possible, to avoid metaphysics.

One point in which language tends to mislead is that the words which symbolize relations are themselves just as substantial as other words. If we say "Cæsar loves Brutus," the word "loves," considered as a physical event, is of exactly the same kind as the words "Cæsar" and "Brutus," but is supposed to mean something of a totally different kind. It follows that the relation of a word to its meaning must be different according to the category to which the meaning belongs. There *is* in the above sentence a relation which is symbolized by a relation, not by a word; this is the three-term relation of love to Cæsar and Brutus. This is symbolized by the order of the words—*i.e.* by a three-term relation. But in order to mention this relation, it is necessary to treat "love" grammatically as a substantive, which tends to confuse the distinction between a substance and a relation. However, it is not very difficult to avoid the false suggestions due to this peculiarity of language, when once the danger of them has been pointed out.

I come now to the second part of our inquiry concerning substance. Assuming that the physical world consists of substances with qualities and relations, are these substances to be taken as permanent bits of matter, or as brief events?

* See *Principia Mathematica*, vol. i., Introduction to second edition.

Common sense holds the former view, though its "things" are only quasi-permanent. But science has found means of resolving "things" into groups of electrons and protons, each of which *may* be quite permanent. As we saw in Part I., there are some who think that an electron and a proton can annihilate each other, so that even they are not quite permanent. But the question of permanence is not the one which most concerns us. The question is: Are electrons and protons part of the ultimate stuff of the world, or are they groups of events, or causal laws of events?

We have already seen that the physical object, as inferred from perception, is a group of events arranged about a centre. There *may* be a substance in the centre, but there can be no reason to think so, since the group of events will produce exactly the same percepts; therefore the substance at the centre, if there is one, is irrelevant to science, and belongs to the realm of mere abstract possibility. If we can reach the same conclusion as regards matter in physics, we have diminished the difficulty involved in building our bridge from perception to physics.

The substitution of space-time for space and time has made it much more natural than formerly to conceive a piece of matter as a group of events. Physics starts, nowadays, from a four-dimensional manifold of events, not, as formerly, from a temporal series of three-dimensional manifolds, connected with each other by the conception of matter in motion. Instead of a permanent piece of matter, we have now the conception of a "world-line," which is a series of events connected with each other in a certain way. The parts of one light-ray are connected with each other in a manner which enables us to consider them as forming, together, one light-ray; but we do not conceive a light-ray as a substance moving with the velocity of light. Just the same kind of connection may be held to constitute the unity of an electron. We have a series of events connected together by causal laws; these may be taken to *be* the electron, since anything further is a rash inference which is theoretically useless.

What is peculiar about a string of events which physics takes as belonging to one electron is a character which is present approximately in the common-sense " thing," a character which I should define as the existence of a first-order differential law connecting successive events along a linear route. That is to say, given an event belonging to an electron at one place in space-time, there will be other events at certain neighbouring regions of space-time, separated from the first and from each other by small time-like intervals, such that, when the intervals are taken small enough, if a, b, c are three such events, and the interval between a and b is equal to that between b and c, then the difference between a and b tends towards equality with the difference between b and c, in certain measurable respects. This is a way of saying that accelerations are always finite—or, where they are not (as perhaps in quantum phenomena), there are other characteristics involved which are subject to a condition analogous to finite acceleration. Let us take first the common-sense " thing." If I watch a moving object, I have a series of percepts which change gradually, both as regards position and as regards qualities—colour, shape, etc. The gradualness of the change is the criterion by which I am led to regard the percepts as all belonging to one " thing." But on a common-sense basis there are exceptions, such as explosions. Science deals with these as rapid, but not instantaneous, changes, and so removes the exceptions. We thus arrive at the conclusion that, given an event x at a time t, there will be closely analogous events at neighbouring times. We may symbolize this by saying that, if there is an event x at time t, there will be, at any neighbouring time $t+dt$, an event:

$$x+f_1(x)dt+f_2(x)dt^2,$$

where $f_1(x)$ is a continuous function of the time, while $f_2(x)$ is determined by the second-order differential equations of physics. The string of events so connected is called one piece of matter. In the case of the sudden changes contemplated by the quantum theory, there is still continuity in everything except spatial position, and the spatial position undergoes a change which is one of a small number of possible changes. Thus in this case also the new occurrences can be causally connected with the old, though the laws of the connection are somewhat different from what they are in the usual case.

Thus the string of events constituting one material unit is distinguished from others by the existence of an intrinsic causal law, though this law is only differential. A light-wave, in this respect, is analogous to a material unit; it differs in the fact that it spreads spherically instead of travelling along a linear route.*

It will be seen that, if a piece of matter is a string of events, the distinction between motion and other continuous changes is not so simple as it seemed. We could form continuous series of events which would not all belong to one piece of matter; therefore the change from one to another would not be a "motion." A "motion" is a string of events connected with each other according to the laws of motion. This might seem like a vicious circle, but in fact it is not. What we assert is: Strings of events exist which are connected with each other according to the laws of motion; one such string is called one piece of matter, and the transition from one event in the string to another is called a motion. This contains as much as can be verifiable in physics, since every percept is an event. There is no mathematical advantage in asserting more, and to assert more is to go beyond the evidence. Therefore it is prudent, in physics, to regard an electron as a group of events con-

* The non-substantial character of the electron emerges even more forcibly from the Heisenberg theory mentioned in Chapter IV. than from the older theory.

nected together in a certain way. An electron *may* be a " thing," but it is absolutely impossible to obtain any evidence for or against this possibility, which is scientifically unimportant, because the group of events has all the requisite properties.

The light thrown on the notion of substance by the connection between physics and perception, which was the third branch of our problem, has already been touched upon. We saw in former chapters that the physical object to be inferred from perception is a group of events, rather than a single " thing." Percepts are always events, and common sense is rash when it refers them to " things " with changing states. There is therefore every reason, from the standpoint of perception, to desire an interpretation of physics which dispenses with permanent substance. As we have seen that such an interpretation is possible, we shall henceforth adopt it.

There is, however, a view not uncommon in philosophy, and perhaps nearer to common sense than the view which I have adopted. This view is, I think, that of Dr Whitehead. It holds that the different events which constitute a group—whether those which make up a physical object at one time or those which make up the history of a physical object—are not *logically* self-subsistent, but are mere " aspects," implying other aspects in some sense which is not merely causal or inductively derived from observed correlations. I consider this view impossible on purely logical grounds, and have so argued elsewhere. But at the moment I prefer to argue that it is empirically useless. Given a group of events, the evidence that they are " aspects " of one " thing " must be inductive evidence derived from perception, and must be exactly the same as the evidence upon which we have relied in collecting them into causal groups. The supposed logical implications, if they exist, cannot be discovered by logic, but only by observation; no one, by mere reasoning, could avoid being deceived by the three-card trick. Moreover, in calling two events " aspects " of one " thing," we imply that their likeness is more important than their difference; but for science both are facts, and of exactly the same importance. One may say that the theory of relativity has grown up by paying attention to small differences between "aspects." I conclude, therefore, that the " thing " with " aspects " is as useless as permanent substance, and represents an inference which is as unwarrantable as it is unnecessary.

CHAPTER XXIV

IMPORTANCE OF STRUCTURE IN SCIENTIFIC INFERENCE

THE inference from perception to physics, which we have been considering, is one which depends upon certain postulates, the chief of which, apart from induction, is the assumption of a certain similarity of structure between cause and effect where both are complex. I want, in this chapter, to inquire more closely into this postulate, not with a view to establishing its validity, which I shall take for granted, but with a view to discovering what it asserts and what are its consequences.

The first point is to be clear as to what we mean by structure. The notion is not applicable to classes, but only to relations or systems of relations. It is fully defined, and made the basis of a general kind of arithmetic, in *Principia Mathematica*.* But as the later parts of that book are not read, I may be excused for repeating, in outline, what is needed for our present purposes.

Two relations P, Q are said to be "similar" if there is a one-one relation between the terms of their fields, which is such that, whenever two terms have the relation P, their correlates have the relation Q, and vice versa. The most familiar example is that of series: two series are similar when their terms can be correlated without change of order. But it would be a great mistake to suppose that series are the only important application of the notion of similarity between relations. A map, for example, if accurate, is similar to the region which it maps. A book spelt phonetically is similar to the sounds produced when it is read aloud. A gramophone record is similar to the music which it produces. And so on.

* Vol. ii., part iv., *150 ff.

It should be observed that similarity applies not only to two-term relations, but to relations with any number of terms. Suppose we have two relations R, R', each n-adic; suppose there is a one-one relation S which relates all the terms in the field of R to all the terms in the field of R'; let $x_1, x_2, \ldots x_n$ be n terms which have the relation R, and let $x_1', x_2', \ldots x_n'$ be the terms correlated with them by the relation S. Then R and R' are similar if there is a one-one relation S such that, when the above conditions are fulfilled, $x_1', x_2', \ldots x_n'$ have the relation R', and conversely.

Two relations which are similar have the same " structure " or " relation-number." The " relation-number " of a relation is the same as its " structure," and is defined as the class of all relations similar to the given relation. Relation-numbers satisfy all the formal laws of arithmetic which are satisfied by transfinite ordinal numbers; ordinal numbers, both finite and transfinite, are a particular kind of relation-numbers—namely, the relation-numbers of relations which generate well-ordered series.

The formal laws satisfied by relation-numbers are:

$$(\alpha+\beta)+\gamma=\alpha+(\beta+\gamma)$$
$$(\alpha\times\beta)\times\gamma=\alpha\times(\beta\times\gamma)$$
$$(\beta+\gamma)\times\alpha=(\beta\times\alpha)+(\gamma\times\alpha)$$
$$\alpha^{\beta}\times\alpha^{\gamma}=\alpha^{\beta+\gamma}$$
$$(\alpha^{\beta})^{\gamma}=\alpha^{\beta\times\gamma}.$$

They do not in general satisfy the commutative law, nor the other form of the distributive law, viz.:

$$\alpha\times(\beta+\gamma)=(\alpha\times\beta)+(\alpha\times\gamma),\ \text{nor}\ \alpha^{\gamma}\times\beta^{\gamma}=(\alpha\times\beta)^{\gamma}.$$

Relation-numbers are important for the following reason. In addition to the propositions which can be *proved* by logic (considered in Chapter XVII.), there are other propositions which can be *enunciated* by logic, though they cannot be proved or disproved except by empirical evidence. Such, for example, is the proposition: " There are classes which are not finite." This is a proposition which is purely logical in content, but there is no *a priori* way of knowing whether it is true or false. (Many such have been proposed, but they are all fallacious.) Then, again, there are propositions which contain some particular constituent, but would be capable of enunciation in logical terms if that constituent were turned into a variable. Take, *e.g.*: "*Before* is a transitive relation." This is not a statement which pure logic can enunciate, because *before* is an empirical relation. But "R is a transitive relation," where R is variable, can be enunciated by pure logic. We will say that a proposition containing a certain constituent a attributes a "logical property" to a if, when a is replaced by a variable x, the result is a propositional function which can be expressed by logic. The test of a logical property is very simple: apart from the constant a, there must be no constants involved—except such purely formal constants as "incompatibility" and "for all values of x," which are not constituents of the propositions in whose verbal or symbolic expression they occur. It will be seen that transitiveness, *e.g.*, is a logical property of a relation; so is asymmetry or symmetry; so is having n terms in its field; so is, in the case of a three-term relation (*between*), the property of generating a Euclidean space; so is, in the case of a four-term relation (*separation of couples*), the property of generating a projective space; and so on. We can now state the proposition on account of which structure is important.

When two relations have the same structure (or relation-number), all their logical properties are identical.

Logical properties include all those which can be expressed in mathematical terms. Moreover, the inferences from perceptions to their causes, assuming such inferences to be valid, are concerned mainly, if not exclusively, with logical properties. This latter proposition is one which we must now examine.

Take first the relation between the space of physics and the space of perception. Within the private space of one percipient, there is a distinction between perceived space-relations and inferred ones. There is a space into which all the percepts of one person fit, but this is a constructed space, the construction being achieved during the first months of life. But there are also perceived space-relations, most obviously among visual percepts. These space-relations are not identical with those which physics assumes among the corresponding physical objects, but they have a certain kind of correspondence with those relations. If we represent the position, for physics, of visible objects by polar co-ordinates, taking the percipient as origin, the two angular co-ordinates correspond to perceived relations among visual percepts, while the radius vector (except possibly for very small distances) is inferred by means of causal laws. Let us confine ourselves to the angular co-ordinates. My point is that the relations which physics assumes in assigning angular co-ordinates are not identical with those which we perceive in the visual field, but merely correspond with them in a manner which preserves their logical (mathematical) properties. This follows from the assumption that any difference between two simultaneous percepts implies a correlative difference in their stimuli. Consequently, assuming that light travels in straight lines, two objects which produce percepts which differ in perceived direction must differ in some respect which corresponds with perceived direction. But we need not assume that physical direction has anything in common with visual direction except the logical properties implied by the above assumption. I shall, in Part III., attempt a construction of physical space which will supply some of the detail of the correspondence; for the present, I am concerned to point out that we can only infer the logical (or mathematical) properties of physical space, and must not suppose that it is identical with the space of our perceptions. Indeed, as I shall try to prove later, the whole of a man's visual space is, for physics, inside his head; this will follow from causal considerations.

The same sort of considerations apply to colours and sounds. Colours and sounds can be arranged in an order with respect to several characteristics; we have a right to assume that their stimuli can be arranged in an order with respect to corresponding characteristics, but this, by itself, determines only certain logical properties of the stimuli. This applies to all varieties of percepts, and accounts for the fact that our knowledge of physics is mathematical: it is mathematical because no non-mathematical properties of the physical world can be inferred from perception.

There is, however, one exception to this limitation, at least apparently. The exception I mean is *time*. We always assume that the time between percepts is the same as the time in the physical world. I do not know whether this view is correct or not; but I will try to set forth the arguments on either side.

In the first place, we must adapt our language to the theory of relativity. I shall assume (what I shall argue in Part III.) that, when we are speaking of physical space, all our percepts are in our head. Consequently psychological time is the same as time measured by our watches, assuming that we carry them on our person. Our head moves along a world-line, and our psychological time-intervals are measured physically by integrating ds along this world-line. Thus there is no difficulty in adapting the statement that psychological and physical time are identical to the requirements of the theory of relativity. In this respect, time differs from space, because physically all our simultaneous percepts are in one place.

I think, however, that the time-intervals between percepts are only to be obtained by means of inferences of the same sort as those which lead us to the physical world. *Perceived* relations are not between events at different times, but between a percept and a recollection, both of which occur at the same time; or again, where very short times are concerned, between a sensation of maximum vividness and a fading (akoluthic) sensation. Sensations do not decay suddenly, but fade gradually, though very quickly. That is why a quick movement can be apprehended as a whole: the sensations belonging to earlier parts are still present, though less vivid, when the sensations belonging to later parts arise. Thus our knowledge of time seems to be inferred from perceived relations which are not strictly temporal. These relations are, I think, of three sorts. Two sorts have been mentioned: the relation of a vivid to a fading sensation, and the relation of a percept to a recollection. But in addition to these there is an order within recollections: we can recollect a process in the right order. Here, also, however, all that we perceive is in the present, and the time-order of the original events is inferred from relations among the simultaneous events which constitute our present recollection. Thus the conclusion seems to be: Psychological time may be identified with physical time, because neither is a datum, but each is derived from data by inferences of the sort we have found elsewhere, namely, inferences which allow us to know only the logical or mathematical properties of what we infer.

Thus it would seem that, wherever we infer from perceptions, it is only structure that we can validly infer; and structure is what can be expressed by mathematical logic, which includes mathematics.

Before concluding this discussion, we must consider an extension of the notion of similarity which has considerable importance in relation to the inferences leading to the physical world. In defining similarity, we used a one-one relation S. But we may substitute a many-one relation, and still obtain something useful. The importance of this is that, as we have seen, if we take a group of events constituting a physical object, the relation of the events which are nearer the object to those which are further from it is many-one, not one-one. If we are observing a man half a mile away, his appearance is not changed if he frowns, whereas it is changed for a man observing him from a distance of three feet. Considerable events may happen in the sun without being perceptible to us even with the best telescopes; but near the sun they may have effects which would be important to a percipient situated where these effects occur. It is obvious as a matter of logic that, if our correlating relation S is many-one, not one-one, logical inference in the sense in which S goes is just as feasible as before, but logical inference in the opposite sense is more difficult. That is why we assume that differing percepts have differing stimuli, but indistinguishable percepts need not have exactly similar stimuli. If we have xSx' and ySy', where S is many-one, and if y and y' differ, we can infer that x and x' differ; but if y and y' do not differ, we cannot infer that x and x' do not differ. We find often that indistinguishable percepts are followed by different effects—*e.g.* one glass of water causes typhoid and another does not. In such cases we assume imperceptible differences—which the microscope may render perceptible. But where there is no discoverable difference in the effects, we can still not be sure there is not a difference in the stimuli which may become relevant at some later stage.

When the relation S is many-one, we shall say that the two systems which it correlates are "semi-similar."

This consideration makes all physical inference more or less precarious. We can construct theories which fit the known facts, but we can never be sure that other theories would not fit them equally well. This is an essential limitation on scientific inference, which is generally recognized by men of science: no prudent man of science would maintain that such-and-such a theory is so firmly established that it will never call for modification. Newtonian gravitation came nearer to this certainty than any other theory has ever done; yet Newtonian gravitation has had to be modified. The fundamental reason for this uncertainty, which remains even when we assume all the canons of scientific inference, is the fact that our relation *S*, which connects the physical object with the percept, is many-one and not one-one.

CHAPTER XXV

PERCEPTION FROM THE STANDPOINT OF PHYSICS

HITHERTO we have been taking perception as our starting-point, and considering how physics could be obtained as an inference from perception. In the present chapter, I want to pursue the opposite course, and consider how, assuming physics, percepts can find their place in the physical world.

Let us first of all exclude certain problems which are not relevant to this inquiry. A "percept," considered as the epistemological basis of physics, must be a "datum"—it must be something noticed. Obviously, therefore, whatever may be true of percepts in general, those which afford empirical premisses for physics have to be "known." But it is unnecessary for us to define "knowing": for physics, only the percepts are important, and our relation to them may be taken for granted. Similarly we need not consider whether, when we perceive, the occurrence is relational, involving a percept and a percipient, or whether the occurrence of the percept is all that happens at the moment, and its "mental" character is conferred by memory (in its most general sense). Such psychological questions need not concern us. What I wish to discuss is the physical status of percepts, *i.e.* of patches of colour, noises, smells, hardnesses, etc., as well as perceived spatial relations. And in this discussion I am now assuming ordinary physics, subject to the latitude of interpretation explained in Chapter I.

Dr Whitehead's books are a protest against the "bifurcation of nature" which has resulted from the causal theory of perception. With this protest I am in complete agreement. Locke's belief, that the primary qualities belong to the object and the secondary to the percipient, has been that of science in practice, whatever individual scientific men may have thought in their philosophic moments. The view which I wish to advocate is quite different. I hold that the world is very full of events, that often a group of these events, or some characteristic which the members of the group possess in varying degrees, is such as to suggest arrangement in an order, generally a symmetrical order about a centre—*e.g.* the percepts of different people when they look at a penny may be ordered by their size and by their shape. The orders derived from different sources are roughly identical: *e.g.* if we move so as to make the big drum look larger, we also move so as to make it sound louder. In this way we construct a space containing both percipients and physical objects; but percepts have a twofold location in this space, namely that of the percipient and that of the physical object. Keeping one half of this location fixed, we obtain the view of the world from a given place; keeping the other half fixed, we obtain the views of a given physical object from different places. The first of these *is* a percipient, the second *is* a physical object. But the first half of this statement is to be taken with a grain of salt.

The physical world, I suggest, considered as perceptible, consists of occurrences having this twofold location. For the moment I am concerned to assign the place of perception in such a scheme.

Consider a spherical light-wave proceeding from a momentary flash. *In vacuo*, it advances in accordance with Maxwell's equations, but when it encounters matter it becomes transformed in one way or another according to circumstances. What do I mean by saying that it "encounters matter"? The answer is quite straightforward. Connected with each electron or proton there is a gravitational field and an electromagnetic field; these are displayed by laws modifying the "undisturbed" distribution about other centres of such things as light-waves. In fact, the fields may be said actually to consist of the formulæ of such modification. Therefore when I say that a light-wave " encounters matter," I mean that it is near the centre of some such systematic modification. The eye is a collection of such centres, and after traversing it the process which was a light-wave obeys a different set of laws. The percept is a term of this process, characterized by the fact that it occurs after traversing a region of a certain sort —to wit, an eye, an optic nerve, and part of a brain. Owing to its causal continuity with other parts of the process, it has, as its twofold location, on the one hand the source of light, on the other hand the brain. If it is said that a percept is " obviously " not in the brain, that is because we are thinking of its location in the physical object, and comparing this with the location of the brain as a physical object.

Certain explanations are called for, chiefly in virtue of Dr Broad's criticisms.* In the first place, it is suggested that the above theory takes a common-sense view of the percipient's body, and derives from this an undue plausibility for the view which it suggests as to external objects. This is not the case, but in order to dispel the appearance of such an error it is necessary to explain the twofold character of a physical object. On the one hand, it is a group of " appearances "—*i.e.* of connected events—differing, from next to next, approximately according to the laws of perspective. On the other hand, a physical object has an influence upon the appearances of other objects, especially appearances in its neighbourhood, causing these to depart, in a greater or less degree, from what they would be if they followed the laws of perspective strictly. The sense organs have only this second function to perform in the theory of perception, while the object perceived has the first function. It is this difference of function, in the theory of perception, which makes it seem as if we were treating the

* *Scientific Thought*, Kegan Paul, 1923, pp. 531 ff., esp. p. 533.

percipient's body more realistically than external objects. But this is only a matter of degree. The appearance of an external object is modified also by other external objects—*e.g.* by blue spectacles or by a microscope. I conceive the part played by the eye as essentially analogous to that played by a microscope; and I take the same view as to the part played by the optic nerve.

Another objection urged by Dr Broad is that the above theory is at best only suitable to visual objects, not to objects known by other senses. Now I certainly hold that vision is much the most important and least misleading of the senses, when considered as a source of the fundamental notions of physics. But I do not admit that the view which I have suggested is in any way inapplicable to the other senses. This subject, however, demands some discussion.

Let us take first the sense of touch. This sense is complicated by the fact that it has no special organ, such as the eye, but is diffused throughout the surface of the body. In order to avoid complications, let us assume that only the tip of the forefinger of the right hand is being used. I do not know what, exactly, is supposed to be the physical process in touch, but we may suppose that it is somewhat as follows: the electrons and protons of a certain part of the skin come into such close proximity to those of an external body that electrical disturbances are set up, which travel along the afferent nerves to the proper part of the brain, and produce corresponding disturbances there. It does not matter for our purposes if this view is not quite right, since the exact nature of the process is irrelevant. But there is one point of some importance, and that is, that the change or lack of change in a sensation of touch has more importance than in the case of sight. A printed letter, and even a printed word, can be seen at a glance; but to read "Braille" it is necessary to let the finger travel round the contours of the letters. Thus shape, in the case of touch, is, in the main, inferred by means of movement; the momentary datum is much simpler than many visual data. The inference to shape depends, of course, upon the assumption that the object touched has not changed its shape meanwhile; it would be difficult for a blind man to acquire correct views as to the shape of an eel. But when there is doubt the finger can be allowed to travel repeatedly round the contours of the object; if the result is similar on each occasion, it may be assumed that the object has kept an approximately unchanging shape.

There is another respect in which touch is inferior to sight, and that is, that the spatial relation of the physical object to the percipient's body is much more restricted. The physical object must be very close to the part of the percipient's body which is said to be touching it. This means that its location is confined within a certain small region. Within that region touch can locate it rather well, provided a sensitive part of the skin is used; we know the position of our hand by means of feelings connected with the muscles, and thence we know the position of anything in contact with the hand. The intervening medium, in the case of touch, is always a part of the percipient's body; but its influence is shown in the difference between the touch sensations when a physical object touches one part of the body and when it touches another. Thus our theory applies to touch just as well as to sight.

Sound is, in many ways, very analogous to light. It is a disturbance having a centre, and is greatest near the centre. What we hear is loudest when we are near the centre. The direction of the sound can be gauged roughly, though not with anything approaching the precision with which we can gauge the direction of a visual object. Here, also, we have a certain physical process, which obeys certain laws in air, but obeys somewhat different laws in the ear and nerves and brain. These differences, however, may be conceived to be of the same kind, essentially, as those normally produced in physical processes by the presence of matter. I cannot see, therefore, that sound offers any difficulty.

The other senses are much less important as sources of physical knowledge, and it seems unnecessary to discuss them in detail. Physiology, however, tends to show that any abnormal condition of the sense organs or of the afferent nerves tends to modify percepts in such a way as requires, for its explanation, some such theory as ours. It is a fallacy to argue, as is sometimes done, that, if we cannot trust our senses, we cannot know that we have sense organs, or that there is any truth in physiology. If we find that several people, looking at Jones, see him just as usual, while one person sees him looking queer; if the several see nothing queer in each other's eyes, while they all see something queer in the eyes of the one; in such circumstances, I say, it is natural and proper to correlate the two queernesses. The man who sees Jones differently from usual sees him through a medium which has an unusual effect; there is no more ground for scepticism than is to be derived from the effect of opera glasses. The sceptical argument is only valid as against naive realism, and derives its rhetorical force from our tendency to relapse into naive realism whenever we are not on our guard.

The cognitive efficacy of perception depends upon two factors, one physical and one psychological (and physiological). The psychological factor is memory and the whole effect of experience upon mind and body. This is a large subject, which I mention only to dismiss. The physical factor, however, may be pointed out once more. It is, the fact that physical occurrences tend to be grouped about centres, the members of one group being approximately related according to laws which we have called the laws of perspective. This enables us to infer from a percept other percepts which we should have if we moved, or which other percipients have now.

When one astronomer sees an eclipse of the moon, he can be pretty sure that others see it too if they are looking in the right direction. When one man sees the Derby, he can be pretty sure that the other spectators are also seeing it—*i.e.* that they have percepts which can be inferred approximately from his by the laws of perspective. As to what is happening where there is no percipient, we can, on certain assumptions, infer a good deal as to its mathematical structure, but nothing as to its intrinsic quality. In a word, the inferential power of perception depends upon the fact that physical events occur in connected groups, and is limited by the fact that this is only true to a certain degree of approximation.

There remains one matter of considerable importance to be discussed in this connection—I mean, the *prima facie* difference between a percept and a physical process. At first sight, a light-wave seems very different from a visual percept, and a sound-wave from an auditory percept. But this apparent gulf is due to comparison of events of different orders. A physical disturbance, such as a light-wave, must be regarded as much more complex in reality than in mathematics. Events in the physical world are correlated according to certain laws, and we can, for mathematical purposes, treat a whole group of correlated events as if it were one event. There is no theoretical reason why a light-wave should not consist of groups of occurrences, each containing a member more or less analogous to a minute part of a visual percept. We cannot perceive a light-wave, since the interposition of an eye and brain stops it. We know, therefore, only its abstract mathematical properties. Such properties may belong to groups composed of any kind of material. To assert that the material *must* be very different from percepts is to assume that we know a great deal more than we do in fact know of the intrinsic character of physical events. If there is any advantage in supposing that the light-wave, the process in the eye, and the process in the optic nerve, contain events qualitatively continuous with the final visual percept, nothing that we know of the physical world can be used to disprove the supposition.

The gulf between percepts and physics is not a gulf as regards intrinsic quality, for we know nothing of the intrinsic quality of the physical world, and therefore do not know whether it is, or is not, very different from that of percepts. The gulf is as to what we know about the two realms. We know the quality of percepts, but we do not know their laws so well as we could wish. We know the laws of the physical world, in so far as these are mathematical, pretty well, but we know nothing else about it. If there is any intellectual difficulty in supposing that the physical world is intrinsically quite unlike that of percepts, this is a reason for supposing that there is not this complete unlikeness. And there is a certain ground for such a view, in the fact that percepts are part of the physical world, and are the only part that we can know without the help of rather elaborate and difficult inferences.

CHAPTER XXVI

NON-MENTAL ANALOGUES TO PERCEPTION

As we saw in Chapter XXV., the cognitive value of perception —*i.e.* its capacity for giving rise to inferences which are often valid—is a product of two factors, one depending upon the human mind and body, the other purely physical. The factor which depends upon the human mind and body is that which is concerned with "mnemic" phenomena. These occur wherever there is life, and to some slight extent in "dead" matter; but the higher the type of life the more notable they become. It is, however, the physical factor in perception that I wish to consider in this chapter, as it appears when separated from the mnemic factor. That is to say, I want to emphasize the fact that a percept is one of a system of correlated events, all structurally similar or semi-similar, and that the physical world, so far as known, consists of such events. My main purpose in dwelling upon this topic is to make it clear that percepts fit easily and naturally into their place in the physical world, and are not to be regarded as something quite different from the processes with which physics is concerned.

Let us revert to our earlier illustration of the dictaphone and camera which record a conversation with its accompanying action, and are found to agree with the recollections of eye-witnesses. When we considered this coincidence in a previous chapter, we were concerned with fundamental doubts; now we will assume the four-dimensional manifold of physics and the justification (in principle) of the inference from perceived to unperceived events. Assuming this, what can we infer as to the relation between (*a*) the sounds heard by the listener, (*b*) the events just outside his ear when he hears, (*c*) the events at the dictaphone at the same time, (*d*) the dictaphone record, (*e*) the sounds heard by the man when he listens to the dictaphone ?

The similarity between (*a*) and (*e*) is fundamental, and is known by a comparison of a percept with a memory. Thus the problem of the relation between perception and memory is involved; but as this problem is psychological, I will only say that the inference from a recollection (which occurs now) to what is recollected (which occurred at a former time) appears to me to be essentially similar to the inferences in physics, and to warrant only a belief in identity (or close similarity) of structure between the recollection and the event recollected. The grounds for the trustworthiness of memory seem to be of the same kind as those for the trustworthiness of perception. But I shall take all this for granted, since our theme is physics, not psychology. I shall therefore assume that (*a*) and (*e*) can be known to be similar in structure, in the sense explained in Chapter XXIV.

We have thus a chain of processes, (*a*) at one end and (*e*) at the other; the end-processes are similar in the technical sense, and we assume that the intermediate processes are also similar, both to each other and to the end-processes. Let us consider this in somewhat more detail. The relation of (*a*) and (*b*) is that of percept and stimulus—*i.e.* a relation of effect to cause. The effect is a complex process; we assume that recognizably different percepts must have different stimuli; therefore the cause must be a complex process, at least semi-similar to the effect. We may take it as similar, not merely semi-similar, by ignoring those respects, if any, in which the structure of the cause is more complex than the structure of the effect. A similar argument will enable us to treat (*d*) and (*e*) as similar. Since (*a*) and (*e*) are similar, it follows that (*b*) and (*d*) are similar. We cannot attribute this similarity to chance, since it is found to exist whenever the necessary conditions have been fulfilled. Hence we infer that (*c*) must also be similar to the other processes. Since the dictaphone may be placed anywhere in the neighbourhood of the speakers, we infer that throughout a region surrounding them there are physical events similar in structure to the aural percepts of the listener. For light, the same thing follows from photographs. Consequently a percept, considered physically, is not very different from other physical events. We may suppose, if we choose, that it differs from them in intrinsic quality, and we know that it differs causally, since it gives rise to memories and inferences. Even these, however, are not so different from certain physical processes as they seem at first sight.

Memory is shown by the capacity for producing events similar in structure to certain previous events, when the right stimulus is applied. We are not always remembering everything that we can remember; we remember things when we are asked about them, or when something occurs which recalls them by association. The dictaphone "remembers" in this sense. It is true that it cannot "infer": it will not answer a question which it has never heard answered. But physiological inference, which is causally the basis of all other inference, is not very unlike other physical processes, and may quite possibly proceed according to the laws of physics. However, I do not wish to pursue these psychological topics; it is only perception and its non-mental analogues that I wish to consider.

We have to suppose that a great many events are taking place everywhere, since both light and sound can be recorded by instruments and observed by percipients. Our visual field is very complex, and the physical stimulus must have at least equal complexity: if this were not the case, we could not see a number of objects at once, nor could a photographic plate photograph them. Physics, however, simplifies all this by taking the stimulus to a sensation to be a periodic process, not a static event. Our perception of colour, for example, does not *seem* to be a periodic process analogous to a light-wave; in this respect, the apparent structure of a visual percept differs from that which physics assumes in the external cause. A few words must be said on this topic, in order to make clear its relation to our general theory of similarity of structure.

First: in a transaction such as the passage from stimulus to percept, we cannot expect *complete* similarity of structure: at most we can expect as much as we find in purely physical transactions. There is a great deal of difference between a light-wave and a quantum change in an atom, yet they are related as effect to cause. What we know about the atom we know in virtue of the light-waves which make us see things; unless differences in light-waves corresponded to differences in atoms, light-waves would not be vehicles of information about atoms. Now when light-waves reach the eye, they have effects upon the matter of the eye, which reverse the previous process from quantum changes to light-waves. It is possible, in view of such theories as we considered in Chapter XIII., that the relation between what happens in the atom and what happens in the eye is more direct than the above account would suggest, but it would not be prudent to assume that this is the case until the theory of light quanta has become more adequate. We cannot, therefore, assume any very close relation between the physical process in the eye and the physical process in the atom from which the light comes. And *a fortiori* we cannot assume a very close relation between the percept and the process in the radiating atom. Yet it is only in so far as such a relation exists that vision can be accepted as a source of physical knowledge; in so far as the correspondence fails, vision ceases to be trustworthy.

Secondly: there is no reason why the degree of correspondence between stimulus and percept which is required should not exist between a periodic process and a static occurrence. So long as different processes give rise to different percepts, the requisites in the way of correspondence are satisfied. There is therefore no theoretic difficulty in the view that the stimulus to a sensation of red is a vibration, while the sensation of red itself has not this character, but is a steady state capable of continuing for a short finite time.

Thirdly: we do not really know that our percept of a colour does not have the rhythmic character of the stimulus. We know something about percepts, but not all about them. We all know that if an object is made to rotate rapidly, for instance on a top, we can see it rotating if it does not go too fast, but when it passes a certain speed we see only a continuous band. This is to be expected in view of the existence of akoluthic sensations. But it by no means follows that there is not a flicker in the percept, although we cannot perceive a flicker. Exactly the same thing applies to light and sound generally, and to the apparent continuity of motion in the cinema. We cannot know, unless in virtue of some elaborate argument, whether our percepts are static or rhythmical, nor yet whether their physical stimuli are continuous or discrete. Such knowledge is rendered impossible by the fact that we can only assume semi-similarity, not full similarity, between percept and stimulus.

There is therefore no difficulty in the accepted theory that the stimuli to our most important percepts are rapid periodic processes. On the other hand, there is a great advantage in this theory, in that it simplifies the physical world which has to be assumed as the cause of our perceptions. A physical system, conceived merely as a set of material units in space-time, is capable of an indefinite variety of rhythmic movements. Some physical structures are resonant for one period, some for another. Thus our sense-organs can select one sort of movement as the stimulus to which they will respond, and reject all the rest. In fact, it may be said that the essential characteristic of a sense-organ is sensitiveness to one sort of stimulus, which, in the case of the eye or the ear, must be a periodic movement. In this the sense-organs do not differ from lifeless instruments, such as photographic plates and gramophones. Such instruments have something closely analogous to perception, when we leave out of account the mental consequences which we observe in ourselves as a result of perception. And in a certain extended sense we may say that every body which behaves in a characteristic manner when a certain stimulus is present, and only then, has a "perception" of that stimulus. We can infer the stimulus from the behaviour of such a body just as well as from our own percepts—sometimes better, as in the case of a very sensitive photographic plate.

The outcome of the discussion we have been conducting in Part II. has been to justify the ordinary scientific attitude, and to minimize the gulf which seems at first sight to exist between perception and physics. We have seen that the inference from percepts to unperceived physical events, though it cannot be made mathematically cogent, is quite as good as any inductive inference can hope to be. And we have found that there is no ground in philosophy for supposing the physical world to be very different from what physics asserts it to be. But we have found it necessary to emphasize the extremely abstract character of physical knowledge, and the fact that physics leaves open all kinds of possibilities as to the intrinsic character of the world to which its equations apply. There is nothing in physics to prove that the physical world is radically different in character from the mental world. I do not myself believe that the philosophical arguments for the view that all reality must be mental are valid. But I also do not believe that any valid arguments against this view are to be derived from physics. The only legitimate attitude about the physical world seems to be one of complete agnosticism as regards all but its mathematical properties. However, something can be done in the way of constructing possible physical worlds which fulfil the equations of physics and yet resemble rather more closely the world of perception than does the world ordinarily presented in physics. Such constructions have the merit of making the inference from perception to physics seem more reliable, since they save us from the necessity of assuming anything radically different from what we know. From this point of view, they have a certain interest, and I shall partially develop them, at least as regards space-time, in Part III. But they must not be confounded with scientific knowledge: they are hypotheses which may hereafter prove fruitful, and which have already a certain imaginative value. But they are not to be regarded as necessitated by any recognized principle of scientific inference.

PART III

THE STRUCTURE OF THE PHYSICAL WORLD

CHAPTER XXVII

PARTICULARS AND EVENTS

WE shall be concerned, in what follows, with the construction of a map of the physical world, in part more or less conjectural, but never in contradiction to the physical or epistemological results hitherto considered. We shall seek to construct a metaphysic of matter which shall make the gulf between physics and perception as small, and the inferences involved in the causal theory of perception as little dubious, as possible. We do not want the percept to appear mysteriously at the end of a causal chain composed of events of a totally different nature; if we can construct a theory of the physical world which makes its events continuous with perception, we have improved the metaphysical status of physics, even if we cannot prove more than that our theory is possible. In what follows, some portions will be more conjectural than others, but I shall try to indicate, at each stage, whether I am advancing what I believe to be a well-grounded inference by induction and analogy, or whether I am concerned only with an illustrative hypothesis designed to exhibit the possibilities that are compatible with the abstract scientific knowledge to be derived from physics.

We have found, hitherto, that what we know of the physical world falls into two parts: on the one hand, the concrete but disjointed knowledge of percepts; on the other hand, the abstract but systematic knowledge of the physical world as a whole. Certain questions as to structure are answered by physics, while others are left open. The questions which are left open are of a sort of which some must always remain open—namely, Is any further analysis of the terms which are ultimate for physics possible, and, if so, what means exist of conjecturing its nature? In science, we have evidence of structure down to a certain point, while beyond that point we have no evidence. There can never be evidence that the point we have reached is one beyond which there is no structure—*i.e.* that we have arrived at simple units totally devoid of parts; therefore analysis is essentially incapable of reaching a term *known* to be final, even if it has in fact reached a final term. I think that, in the case of physics, there is reason to think that its terms are not final, and that it is possible to suggest a further analysis which is at least likely to be true.

When we wish to describe a structure, we have to do so by means of terms and relations. It may turn out that the terms themselves have a structure, as, *e.g.*, in arithmetic, when cardinal integers are defined as classes of similar classes. In the technique of mathematical physics, there is a considerable apparatus which belongs to the formal method, and would not be regarded by most physicists as having any physical reality. Such is the manifold of space-time points. Space-time is held to represent a system of physical facts, but its mathematical points are generally conceded to be fictions. Such a state of affairs is unsatisfactory until we can say just what non-fictional assertion is implicit in a true proposition of physics which technically uses "points." I propose to deal with this problem in the next chapter.

But what shall we say of electrons? Are they physical realities, or are they mathematical conveniences, like points? Or are they something intermediate between these two extremes? We think of a light-ray as a series of events; is an electron perhaps something similar? But the light-ray also raises problems: it has a certain assigned mathematical structure, but it is difficult to say what we are to think of the mathematical terms of this structure. Formerly, the conception of a transverse wave in the æther seemed fairly clear:

the æther was composed of particles, each of which could move in the required manner. But nowadays the æther is grown insubstantial and incapable of " motion " in any straightforward sense; certainly few people would venture to regard it as composed of point-particles, like the homogeneous fluid of a hydrodynamical text-book. Thus the light-wave has become a structure in the air, like a genealogical tree whose members are all imaginary. This illustrates a necessity in describing a structure: the terms are as important as the relations, and we cannot rest content with terms which we believe to be fictitious. It is the terms of the physical structure that will concern us in the present chapter.

I shall give the name " particulars " to the ultimate terms of the physical structure—ultimate, I mean, in relation to the whole of our present knowledge. A " particular," that is to say, will be something which is concerned in the physical world merely through its qualities or its relations to other things, never through its own structure, if any. The difference between a transverse wave and a longitudinal wave is a difference of structure; therefore neither can be a " particular " in the technical sense in which I mean it. An atom is a structure of electrons and protons; therefore an atom is not a " particular." But when I call something a " particular," I do not mean to assert that it certainly has no structure; I assert only that nothing in the known laws of its behaviour and relations gives us reason to infer a structure. From the standpoint of logic, a particular fulfils the definition of " substance " which we gave in Chapter XXIII. But it fulfils this definition only in the existing state of knowledge; further discoveries may require us to recognize structure within it, and it will then cease to fulfil the definition of substance. This does not falsify former statements as to the structure of the world, in which the particular in question was taken as unanalyzable; it merely adds new propositions, in which it is no longer so treated. Atoms were formerly particulars; now they have ceased to be so. But that has not falsified the chemical propositions which can be enunciated without taking account of their structure. The word "particular," as above defined, is, therefore, a word relative to our knowledge, not an absolute metaphysical term.

Let us begin with a few general considerations as to our knowledge of structure. Part of this knowledge is obtainable by analysis of percepts, part depends upon inferences involving unperceived entities. I shall call a relation "perceived" or "perceptual" if the fact that this relation holds between certain terms can be discovered by mere analysis of percepts. Thus before-and-after is a perceptual relation, when it occurs between terms both of which belong to the specious present. Spatial relations within the visual field are perceptual; so are those between simultaneous tactual sensations in different parts of the body. Tactual sensations in the same part of the body, say a finger-tip, may have perceived relations, if both are within the specious present; these must be important in the recognition of shape by blind people. There are perceived relations between a percept and a recollection, which lead us to refer the latter to the past. There are perceived relations of comparison, which may sometimes be rather complicated—*e.g.* "The resemblance of blue and green is greater than the resemblance of blue and yellow." (Here the blue and green and yellow are supposed to be particular given patches of colour.) There is also, I should say, a perceived relation of simultaneity. I do not suggest that the above list is complete, but it indicates the kinds of cases in which relations can be perceived.

There is a well-advertised type of difficulty in such cases as the analysis of a perceived motion. If I move my hand before my eyes from left to right, and attend to the visual percept, it seems qualitatively different from the successive perceptions of my hand in a number of different positions. On a watch, we can "see" the motion of the second hand, but not of the minute hand. There is no doubt that there is an occurrence which we naturally describe as the perception of a motion. We are aware of perceiving a process: if I move my hand from left to right, the impression is different from what it is if I move my hand from right to left, and it is obvious to everyone that the difference is in the "sense" of the motion. We can, in fact, distinguish earlier and later parts of the motion, so that the motion does not appear to be without structure. But the parts of it seem to be other motions, which, presumably, must each have its own structure. This leads to the notion of infinite divisibility, not based upon a definable structure of indivisibles, but upon a process in which the parts are always composed of parts similar in structure to themselves, and simple parts are nowhere attainable. The paradoxes of motion, the antinomies, Bergson's objection to analysis, and the philosophers' insistence that the Cantorian continuum does not resolve their difficulties, are all derived from this one puzzle, that a motion seems to consist of motions —or, as Kant says, that a space consists of spaces.

It is important to clear up this problem of the analysis of the percept of motion, since it applies to all perception of change, and has been thought to constitute a difficulty in the attempt to harmonize psychology and physics. To begin with, continuity in the percept is no evidence of continuity in the physical process; it is easy to produce a staccato process which causes a continuous (or apparently continuous) percept —*e.g.* in the cinema. Next, it is noteworthy that, if a staccato physical process is gradually accelerated, the percept will retain its staccato character longer if we are wide awake and have acute senses than if we are sleepy or have feeble senses. Everybody knows the experience of being awakened from a doze by a striking clock: at first, the noise of the strike seems continuous. It is therefore a tenable hypothesis, if desirable on other grounds, to maintain that all physical processes are staccato, and continuity in percepts is merely a case of vagueness, in the sense of a many-one relation between stimulus and percept. I am not asserting such a view; I am only saying that it fits in with what we know of the relation between stimulus and percept in the case of swift processes. *A fortiori*, the mathematical continuum, if it existed in the stimulus process, would produce the percepts we call continuous. There is therefore nothing in our perception of process to make us feel that the mathematical analysis of continuity must be inadequate to physics, nor yet to show that a quantized time and space could not produce the sort of percepts which we call "seeing a motion." All physical possibilities are left open, so far as the immediate character of the percept is concerned.

The argument advanced by those who lay stress upon the perceived character of perceptual continuity is, however, not as to the nature of the physical stimulus, but as to the nature of the percept. The continuity of the percept, they maintain, is quite obviously not that of the mathematical continuum, nor yet the deceptive appearance of continuity which would exist if the percept were a rapid staccato process. In saying this, they seem to me to go beyond what the evidence warrants. Consider a case which is analogous in some respects, but not in others—namely, the case of slightly different shades of colour. Suppose we have a series of colours, $A, B, C, D, \ldots$ such that each is sensibly indistinguishable from its neighbour, but not from the rest. That is to say, we can see no difference between A and B or between B and C, but we can see a difference between A and C. We are then compelled to infer a difference between A and B and between B and C, although we cannot perceive any difference. There is no theoretical difficulty in such an inference, for, although A and B and C

are percepts, and the difference between *A* and *C* is a percept, there is no reason why the differences between *A* and *B* and between *B* and *C* should be percepts: the relations between percepts are sometimes percepts and sometimes not. Now, instead of different static shades of colour, let us suppose that we are watching a chameleon gradually changing. We may be quite unable to "see" a process of change, and yet able to know that, after a time, a change has taken place. This will occur if, supposing *A* and *B* to be the shades at the beginning and end of a specious present, *A* and *B* are indistinguishable, while *A* recollected is distinguishable from *C* when *C* occurs. The supposition we have to make about a perceived motion is not quite analogous to this, but has certain points in common with it. Suppose that we are perceiving a motion in a case where we know the physical stimulus to consist of a discrete series, as in the cinema. Let us suppose that n of these stimuli can be comprised within one specious present, and that each produces an element in the percept. Then the percept at one instant consists of n elements x_1, x_2, ... x_n, which are arranged in an order by the degree of fading. Let us suppose that we cannot distinguish x_1 from x_2, nor x_2 from x_3, but that we can distinguish x_1 from x_3. In that case our present percept will be indistinguishable from the percept of a continuous motion. The percept will in fact contain parts that are not processes, but these parts will be imperceptible. The analogy with the case of the colours arises through the existence, in each case, of a series in which differences of neighbouring terms are imperceptible while those of distant terms are perceptible. And it elicits the important principle that a percept may have parts which are not percepts, so that the structure of a percept may be only discoverable by inference. It follows also that we need not assume anything mysterious about the kind of complexity belonging to a percept of motion, but may regard its complexity as of the same kind as that belonging to the stimulus according to mathematical physics.

I wish now to consider the general question: how can we infer structure when it is not perceived? The above discussion of motion involved a particular case of such inference, but now I wish to consider the problem more generally.

For reasons analogous to those which arise in analyzing motion, we are led to the view that all our percepts are composed of imperceptible parts. We can, for instance, perceive a heap of fine powder, and remove the whole heap grain by grain, where at each stage there is no perceptible difference. Our original percept may have had perceptible parts, but these were apparently always complex. It is not strictly necessary to suppose the percepts complex; they might form a series of gradually varying quality. But we may say, in a sense, that the difference of *A* and *C* (supposed perceptible) is compounded of the differences between *A* and *B*, *B* and *C* (supposed imperceptible). Thus we arrive at virtually the same result in regard to qualitative differences as we have otherwise in regard to substantial parts. All such arguments rest ultimately upon the logical premiss that exact similarity is transitive, and the empirical premiss that indistinguishability is not transitive. These two together are the source of much of our inference as regards structure.

There is, however, another source, derived from causal arguments. Two indistinguishable percepts are found to be followed by different results. Inverting the maxim "same cause, same effect," we argue: "Different effects, different causes." Often the difference in the causes becomes perceptible under the microscope; but we assume it in any case. It is this, more than anything else, that has led to the minuteness of the processes inferred by physics. There are noticeable differences in the effects in cases where we know that the difference in the causes, if any, must be very small; we are therefore compelled to attribute to the physical world a structure which is very fine-grained relatively to perception.

It is necessary to consider the very usual form of analysis into diversity of "substance," because, for reasons already given, we cannot regard this form of analysis as ultimate. Let us take the most elementary of scientific examples: the analysis of water into hydrogen and oxygen. We recognize water by a group of characteristic percepts and processes; by another group we recognize hydrogen, and by yet another oxygen. We find that we can—*e.g.* by electrolysis—produce hydrogen and oxygen where formerly there was water; we find that the masses of the two bear a fixed proportion to each other, and add up to the mass of the previous water; we find further that, if we let them come together, water reappears, equal in amount to what was lost by electrolysis. Such facts are interpreted in science by means of the postulate that matter is indestructible. If we accept this postulate, the facts prove that water consists of hydrogen and oxygen. Exactly similar arguments lead us on from atoms to electrons and protons, where, for the present, the process of substantial analysis ceases.

Without questioning the *convenience* of substantial analysis, it may be asked whether it is metaphysically accurate, and even whether, at the stage we have reached, it is adequate to all the needs of physics. We must now examine the arguments on this question.

As regards adequacy for physics: we have already (in Chapter IV.) given a brief account of Heisenberg's theory, which, in effect, resolves the electron into a series of radiations. We have also seen that electrons and protons are not now supposed to be strictly indestructible, but are thought by many to be capable of annihilating each other. Thus the indestructibility of matter is no longer accepted as a universal law of the physical world. With this goes the fact that proper mass is not supposed to be exactly conserved, and that relative mass has been absorbed into energy. Mass was supposed to be "quantity of matter." This certainly could not be said of relative mass, which depends upon the choice of axes and belongs also to light-waves. And if it be said of proper mass, we must conclude that the "quantity of matter" is not quite constant. On all these grounds, persistent units of matter, though still convenient, have no longer the metaphysical status that they were formerly supposed to have.

This conclusion is reinforced by arguments of economy. We perceive events, not substances; that is to say, what we perceive occupies a volume of space-time which is small in all four dimensions, not indefinitely extended in one dimension (time). And what we can primarily infer from percepts, assuming the validity of physics, are groups of events, again not substances. It is a mere linguistic convenience to regard a group of events as states of a "thing," or "substance," or "piece of matter." This inference was originally made on the ground of the logic which philosophers inherited from common sense. But the logic was faulty, and the inference is unnecessary. By defining a "thing" as the group of what would formerly have been its "states," we alter nothing in the detail of physics, and avoid an inference as precarious as it is useless.

What, then, shall we say about the analysis of water into hydrogen and oxygen? We shall say something of this sort: Water has, for common sense, a certain amount of permanence: although puddles dry up, the sea is always there. This permanence, interpreted without the use of "substance," means certain intrinsic causal laws: the behaviour of the sea can, to a considerable extent, be discovered by observing only the sea, without taking account of other things. Similarity on different occasions is the most obvious of these approximate causal laws. But water can change into ice or snow or steam:

here we can observe the gradual transformation, and continuity takes the place of likeness for common sense. In all changes, we find, on examination, that there is some continuity like that between water and ice; we thus trace a causal chain, more or less separable from other causal chains, and having enough intrinsic unity to be regarded as successive states of one "substance." When we throw over "substance," we preserve the causal chain, substituting the unity of a causal process for material identity. Thus the persistence of substance is replaced by the persistence of causal laws, which was, in fact, the criterion by which the supposed material identity was recognized. We thus preserve everything that there was reason to suppose true, and reject only a piece of unfruitful metaphysics.

The analysis of water into hydrogen and oxygen represents, therefore, the analysis of one approximate causal law into two more nearly accurate causal laws. If you infer that where there was water yesterday there is water to-day, you are employing a causal law which is not always correct. If you infer that where there was hydrogen and oxygen there is hydrogen and oxygen (or at least that there is hydrogen and oxygen in places connected by a continuous route with where they were yesterday), you are very unlikely to be wrong, unless the place is in the neighbourhood of Sir Ernest Rutherford It is assumed (what is only partially true at present) that the properties of water can be inferred from those of oxygen and hydrogen together with the manner in which they are combined in the molecules of water. Thus by means of analysis you have obtained causal laws which are at once more true and more powerful than those which common sense could obtain by supposing that all the parts of water were water.

We may say that this is the characteristic merit of analysis as practised in science: it enables us to arrive at a structure such that the properties of the complex can be inferred from those of the parts.* And it enables us to arrive at laws which are permanent, not merely temporary and approximate. This is an ideal, only partially verified as yet; but the degree of verification is abundantly sufficient to justify science in constructing the world out of minute units.

From what has been said about substance, I draw the conclusion that science is concerned with groups of "events," rather than with "things" that have changing "states." This is also the natural conclusion to draw from the substitution of space-time for space and time. The old notion of substance had a certain appropriateness so long as we could believe in one cosmic time and one cosmic space; but it does not fit in so easily when we adopt the four-dimensional space-time framework. I shall therefore assume henceforth that the physical world is to be constructed out of "events," by which I mean practically, as already explained, entities or structures occupying a region of space-time which is small in all four dimensions. "Events" may have a structure, but it is convenient to use the word "event," in the strict sense, to mean something which, if it has a structure, has no space-time structure, *i.e.* it does not have parts which are external to each other in space-time. I do not assume that an event can ever occupy only a point of space-time; the construction of "points" out of finitely extended events will form the subject of the next chapter. Nor do I assign a maximum to the duration of an event, though I hold that any event, in the broad sense, which lasts for more

* Dr C. D. Broad, in *The Mind and its Place in Nature*, lays stress upon what he calls "emergent" properties of complexes—*i.e.* such as cannot be inferred from the properties and relations of the parts. I believe that "emergent" properties represent merely scientific incompleteness, which would not exist in the ideal physics. It is difficult to advance any conclusive argument on either side as to the ultimate character of apparently "emergent" properties, but I think my view is supported by such examples as the explanation of chemistry in terms of physics by means of the Rutherford-Bohr theory of atomic structure.

than about a second can, if it is a percept, be analyzed into a structure of events. But this is a merely empirical fact.

There are certain purely logical principles which are useful in regard to structure. When we are dealing with inferred entities, as to which, as explained in Part II., we know nothing beyond structure, we may be said to know the equations, but not what they mean: so long as they lead to the same results as regards percepts, all interpretations are equally legitimate. Let us take an example. Suppose we have a set of propositions about an electron which we will call E. According to the subject-predicate logic, and according to the view that matter is a substance, there is a certain entity E which is mentioned in all statements about this electron. According to the view which resolves an electron into a series of events, the propositions in question will be differently analyzed. Assuming a certain schematic simplicity, we might set the matter out as follows: there is a certain relation R which sometimes holds between events, and when it holds between x and y, x and y are said to be events in the biography of the same electron. If x belongs to the field of R, " the electron to which x belongs " will mean the relation R with its field limited to terms belonging to the R-family of x; and the R-family of x consists of x together with the terms which have the relation R to x and the terms to which x has the relation R. " This electron " will mean " the electron to which this belongs." " An electron " will mean " a series such that there is an x such that the series is the electron to which x belongs." In order to mention some particular electron, we must be able to mention some event connected with it, *e.g.* the scintillation when it hits a certain screen. Thus, instead of saying "the event z happened to the electron E " we shall say " the event z happened to the electron to which x happened," or, more simply, " z belongs to the R-family of x." The formal properties of the propositional function " z belongs to the R-family of x" (R being constant) are the same as those of "z belongs to the electron E." If we want any two electrons to be mutually exclusive, in the sense that no event can happen to both, we can insure it by assuming that if x has the relation R (or the converse relation) to both y and z, then y belongs to the R-family of z. If we do not want this, we do not make this assumption about R. It is because of the identity in formal properties that the one propositional function can be substituted for the other. Whenever we suggest a new view as to structure, we have to make sure that it does not falsify any of the old formulæ, though it may give them a new interpretation.

Another illustration, more purely logical, may be useful. It seems natural to say that any given shade of colour is a quality, *i.e.* that when we say "this is red," we are saying that "this" has a characteristic which we cannot express otherwise than by a predicate—assuming, for the moment, that "red" stands for just one shade of colour. But although this *may* be the right view, there is no logical necessity for supposing that it is. We might define one shade of colour as "all the coloured surfaces which have exact colour-similarity to a given surface." Thus "this has the colour C" is replaced by "this is one of the class of entities that have exact colour-similarity with x"; and "C is a colour" will be replaced by "C is the class of all entities having exact colour-similarity with a given entity." In this case, no facts can be conceived which would give reason for preferring one form of statement to the other, since any ascertainable fact can be interpreted equally well on either theory.

We have, in fact, something more or less analogous to the arbitrariness of co-ordinates in the general theory of relativity. Provided our symbols have the same interpretation when they apply to percepts, their interpretation elsewhere is arbitrary, since, so long as the formulæ remain the same, the *structure* asserted is the same whatever interpretation we give. Structure, and nothing else, is just what is asserted by formulæ in which the meaning of the terms is unknown, but the purely logical symbols have definite meanings (see Chapter XVII.). Even the purely logical symbols are arbitrary to a certain limited extent, as we saw in the above example of colours. But often, when facts from different regions have to be brought into connection, one interpretation is much simpler than another. Often, also, one interpretation involves less inference than another, and is therefore less likely to be wrong. These are the main motives governing any suggested interpretation of the symbols which occur in mathematical physics.

CHAPTER XXVIII

THE CONSTRUCTION OF POINTS*

THE subject of this chapter is one which has been treated with wonderful ingenuity by Dr Whitehead, to whom is due the whole conception of a method which arrives at "points" as systems of finitely-extended events. In advocating this method, it is not necessary to maintain that mathematical points are *impossible* as simple entities (or "particulars"); all that it is necessary to maintain is that we have no good ground for regarding them as such. What we know about points is that they are useful technically—so useful that we must seek an interpretation of the propositions in which, symbolically, they occur. But there is no ground for denying structure to a point; on the contrary, there are two grounds for assigning structure to a point. One is the familiar argument of Occam's razor: we can make structures having the mathematical properties of points, and to suppose that there are points in any other sense is an inference which is useless to science and not warranted by any principle, logical or scientific. The other argument is much more difficult to state, but the more one studies logical construction the more weight one feels inclined to attach to it. It rests upon a maxim which might be enunciated as a supplement to Occam's razor: "What is logically convenient is likely to be artificial." To me personally, the first example of this maxim was the definition of real numbers. Mathematicians found it convenient to suppose that all series of rationals have limits, while never-

* In this chapter and the next, I owe much to the criticism and suggestions of Mr M. H. A. Newman of St. John's College, Cambridge, who must not, however, be held responsible for their contents; on the contrary, I am convinced that he could construct a much better theory than that which follows.

theless some do not have *rational* limits. They therefore postulated irrational limits, supposed to be homogeneous with the rationals. Although the method of Dedekind cuts was familiar, nobody thought of saying: An irrational *is* a Dedekind cut, or at least its inferior portion. Yet this definition solves all difficulties. We have now first ratios (which cannot be irrational), then segments of the series of ratios. Segments which have a limit are rational, segments which have no limit are irrational. The square root of 2 is the class of ratios whose square is less than 2. Segments of the series of ratios are "real numbers"; the series of real numbers has both Dedekindian and Cantorian continuity. Thus it is mathematically convenient; but its logical structure is more complex than that of the series of ratios. The logical analysis of mathematics affords many examples of this procedure, such as the construction of "ideal" points, lines, and planes alluded to in Chapter XX.

It will be seen that the phrase "what is logically convenient is artificial" does not express what is meant with as much precision as is to be desired. What we mean is this: Given a set of terms having properties which *suggest* certain general mathematical (or logical) properties, but are subject to exceptions in regard to these properties, it is a mistake to postulate other terms, logically homogeneous with the original set, and such as to remove the exceptions; the proper procedure is to look for logical structures composed of the original terms, and such that these structures always have the mathematical properties in question. It will be found that, where the assumption of such properties has proved fruitful, this procedure is usually possible.

Starting from events, there are many ways of reaching points. One is the method adopted by Dr Whitehead, in which we consider "enclosure-series." Speaking roughly, we may say that this method defines a point as all the volumes which contain the point. (The niceties of the method are required to prevent this definition from being circular; also to distinguish a set of volumes having only a point in common from such as have a line or surface in common.) As a piece of logic, this method is faultless. But as a method which aims at starting with the actual constituents of the world it seems to me to have certain defects. Dr Whitehead assumes that every event encloses and is enclosed by other events. There is, therefore, for him, no lower limit or minimum, and no upper limit or maximum, to the size of events. Each of these assumptions demands consideration.

Let us begin with the absence of a lower limit or minimum. Here we are confronted with a question of fact, which might conceivably be decided against Dr Whitehead, but could not conceivably be decided in his favour. The events which we can perceive all have a certain duration, *i.e.* they are simultaneous with events which are not simultaneous with each other. Not only are they all, in this sense, finite, but they are all above an assignable limit. I do not know what is the shortest perceptible event, but this is the sort of question which a psychological laboratory could answer. We have not, therefore, direct empirical evidence that there is no minimum to events. Nor can we have indirect empirical evidence, since a process which proceeds by very small finite differences is sensibly indistinguishable from a continuous process, as the cinema shows. *Per contra,* there might be empirical evidence, as in the quantum theory, that events could not have less than a certain minimum spatio-temporal extent. Dr Whitehead's assumption, therefore, seems rash. At the same time, there is a confusion to be avoided: space-time may be continuous even if there is a lower limit to events. Suppose every elementary event filled a four-dimensional cube, *e.g.* a cubic centimetre lasting for the time that light takes to travel a centimetre; and suppose, conversely, that every such four-dimensional cube was occupied by an event. The space-time of such a world would be continuous, given suitable axioms, although events had a minimum. And, conversely, the absence of a minimum to events does not insure spatio-temporal continuity. The two questions are thus wholly distinct.

I conclude that there is at present no means of knowing whether events have a minimum or not; that there never can be conclusive evidence against their having a minimum; but that conceivably evidence may hereafter be found in favour of a minimum. It remains to consider the question of a maximum.

On the question of a maximum to events, the arguments are rather logical than empirical. In a certain sense, any series of events may be called one event; the Battle of Waterloo, for instance, may count as a single occurrence. But in a complex event of this sort, there are parts which have spatio-temporal and causal relations to each other; no single entity devoid of physical structure persists throughout the whole period. I mean by this that anything simultaneous with everything that happened during the Battle of Waterloo is a complex of parts not all simultaneous with each other. Whether we are to call such a complex an "event" or not is merely a question of words. But if our object is to exhibit the structure of the physical world, it is clear that we must distinguish objects having physical structure from such as are only component parts of such structures. It is therefore convenient to have a word for the latter. The word I shall use is "event." But I shall not go so far as to say that an "event" must have *no* structure. I shall assume only that any structure which it may have is irrelevant both to physics and to psychology; in other words, that its parts, if any, do not have scientifically distinguishable relations to other objects. When the word "event" is used in this sense, it is plain that, so far as our experience goes, no event lasts for more than a few seconds at most. There is no *a priori* reason why this should be the case; it is merely an empirical fact. But I think a phraseology which obscures it can only lead to confusion.

For the above reasons, I am unable to accept Dr Whitehead's construction of points by means of enclosure-series as an adequate solution of the problem which it is designed to solve. This problem is: to discover structures having certain geometrical properties, and composed of the raw material of the physical world.

There is another method, which may be called that of "partial overlapping." In my *Knowledge of the External World*, I applied this method to the definition of instants. It is easy to see that it is adequate for this purpose in psychology, where we have a one-dimensional time-order which remains definite in spite of relativity. But in physics it is the "point-instant" that has to be defined, *i.e.* a completely definite position in space-time, not merely in space or merely in time. Here the method is only applicable with suitable modifications. However, the method must first be explained as applied to the one-dimensional psychological time-series.

We assume that two events may have a relation which I will call "compresence," which means, practically, that they overlap in space-time. Take, for instance, notes played by different instruments in orchestral music: if one is heard beginning before the other has ceased to be heard, the auditory percepts of the hearer have "compresence." If a group of events in one biography are all compresent with each other, there will be some place in space-time which is occupied by all of them. This place will be a "point" if there is no event outside the group which is compresent with all of them. We may therefore define a "point-instant," or simply a "point," in one biography, as a group of events having the following two properties:

(1) Any two members of the group are compresent;
(2) No event outside the group is compresent with every member of the group.

When we pass beyond one dimension, this method is no longer applicable. Take, for example, the three circles in the accompanying figure: each overlaps with the other two, but there is no region common to all three. If we try to remedy this (as I believe we can) by starting, in two dimensions, with a relation of *three* events, which is to hold when all three have a region in common, we are still met by difficulties. The three circles *a*, *b*, *c* have a region in common, and the shaded area *d* has a region in common with *a* and *b*, also with *a* and *c*, and also with *b* and *c*, yet *a*, *b*, *c* and *d* have no region in common. Therefore if events may have queer shapes such as *d*, our new three-term relation will still not enable us to define a "point."

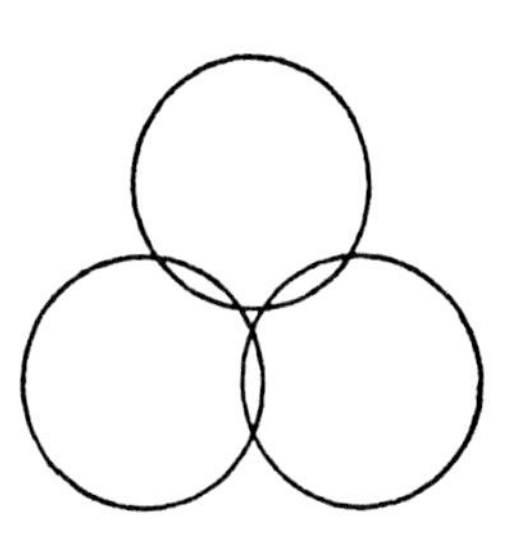

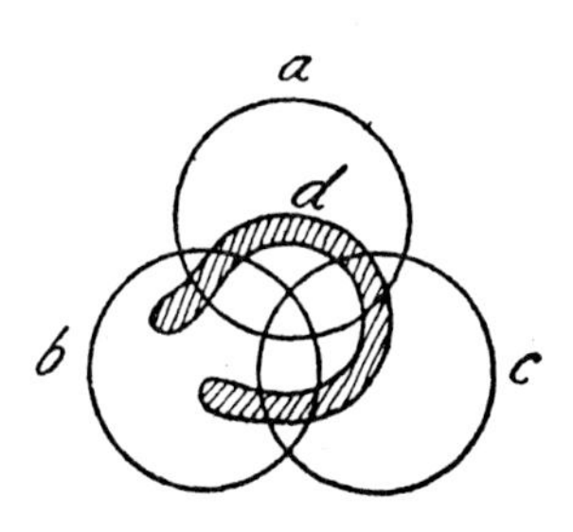

Since the problem with which we are concerned belongs to *analysis situs*, in which we are occupied only with such properties of figures as are unaffected by continuous deformation, we cannot simply declare in advance that no events are to have odd shapes. But before attempting to deal with this difficulty, it will be as well to consider certain points in *analysis situs*, which will show us what are the requisites of a solution of our problem. In *analysis situs* we start with two conceptions, that of a point, and that of "neighbourhoods of a given point"—the latter being collections of points. Certain definitions obtained in this way will be useful.

The following definitions are due to Leopold Vietoris.*

If M is a set of points, a point p is called a "Häufungspunkt" of M if in every neighbourhood of p there is a point other than p.

Two collections of points "touch" each other in a point p if p belongs to one collection and is a "Häufungspunkt" of the other.

A set of points M is "continuous from a to b" if it contains a and b, and any two parts of it whose sum is M, of which one contains a and the other b, touch each other (in at least one point).

A set of points M is a "Linienstück" from a to b if it, but none of its proper parts, is continuous from a to b.

Hausdorff † has defined a "metrical" space and a "topological" space in the following terms.

A "metrical" space is a manifold such that with any two points x, y is associated a real not-negative number xy having the following three properties: (a) $yx = xy$; (b) xy is only zero when x and y are identical; (c) $xy + yz$ is greater than or equal to xz.‡

A "topological" space is a manifold whose elements x are associated with sub-classes U_x of the manifold such that:

(A) To every x corresponds at least one U_x, and every U_x contains x;

(B) If U_x, V_x are both neighbourhoods of x, there is a neighbourhood of x, say W_x, which is contained in the common part of U_x and V_x;

(C) If y is a member of U_x, there is a neighbourhood of y which is contained in U_x;

(D) Given any two distinct points, there is a neighbourhood of the one and there is a neighbourhood of the other such that the two have no common point.§

* *Stetige Mengen*, Monatshafte für Mathematik u. Physik., xxxi., 1921, pp. 173-204.

† *Grundzüge der Mengenlehre*, Leipzig, 1914.

‡ *Ib.*, p. 211.

§ *Ib.*, p. 213.

In order to be able to apply the usual methods of limits to a topological space, Hausdorff has need of an "Abzählbarkeits-axiom," or "denumerative axiom." He gives two such axioms (p. 263), of which the first is the weaker, and is for some purposes insufficient. The first states that the number of neighbourhoods of a given point is never greater than $\aleph_0$; the second states that the total number of neighbourhoods of all points is together $\aleph_0$. This second axiom suffices for all the usual kinds of argument, without the introduction of any metrical ideas.

P. Urysohn* has shown that every topological space which satisfies Hausdorff's second denumerative axiom and has one further property (which he calls "normality"†) is metricizable.

These are the main points from *analysis situs* that are relevant to the solution of our problem.

For the present, we are not concerned with metrical properties, but only with such as belong to "topological" spaces. In virtue of Urysohn's theorem, it will be possible to introduce a metric if we can construct the right sort of topological space. But when one metric is possible, an infinite number are possible. The metric which is actually introduced in theory of relativity is introduced for empirical reasons; it uses a quantitative relation which might be called degree of causal proximity. The existence of this relation is not implied by anything with which we are at present concerned. Moreover, the metrical manifold which we require in physics is not a "metrical space" according to Hausdorff's definition given above, since interval in relativity does not possess the properties (*b*) and (*c*)

* *Zum Metrisationsproblem*, Math. Annalen 94 (1925), pp. 309-315.

† He defines a topological space as "normal" when any two non-overlapping closed manifolds A and B can be separated by two non-overlapping regions G_A, G_B which respectively contain them and have no boundary-points. *Ib.*, p. 310, and Hausdorff, *op. cit.*, p. 215. A "boundary-point" of a collection is one which has a neighbourhood that is not a sub-class of the collection.

which distance possesses in Hausdorff's definition. However, so far as topological considerations are concerned, we may, without appreciable inaccuracy, assign to small regions the topological properties which belong to a small region of Euclidean space lasting for a short time, *i.e.* to a continuous series of small regions of Euclidean space all geometrically indistinguishable.

In *analysis situs*, both points and neighbourhoods are given. We, on the other hand, wish to define our points in terms of "events," where "events" will have a one-one correspondence with certain neighbourhoods. We want our "events" to correspond with neighbourhoods which are above a certain minimum and below a certain maximum when, at a later stage, the empirical metric is introduced. We have to assign to our events such properties as will enable us to define the points of a topological space as classes of events, and the neighbourhoods of the points as classes of points. But we have to remember that we do not want to construct merely a topological space: what we want to construct is the four-dimensional space-time of the general theory of relativity.

The following illustration will serve to introduce the problem. Consider a three-dimensional Euclidean numerical space, *i.e.* the manifold of all ordered triads of real numbers (x, y, z), with the usual definition of distance. Consider, in this space, all the spheres having a given radius and having centres whose co-ordinates are rational. The number of such spheres is $\aleph_0$. Let us define a group of these spheres as "co-punctual" if it is such that every four chosen out of the group have a common region; and let us define a co-punctual group as "punctual" if it cannot be enlarged without ceasing to be co-punctual. Then there is a one-one correspondence between the original points of our space and the punctual groups of spheres. Consequently the punctual groups of spheres form a Euclidean space. If the spheres are all distorted in any continuous way, they will still enable us to construct punctual groups in the same way, and the manifold of punctual groups will still have all the topological properties which are possessed by a three-dimensional Euclidean space. Therefore if we are to use this method of constructing points out of "events," we shall have to assume that, in the resulting space, there is a possible metric according to which the points of which a given event is a member always form a spherical volume. Although this is expressed in metrical language, it is in reality a topological property, since it is unaffected by continuous deformation. It must be possible to express it in non-metrical language, though I must confess that I lack the necessary skill.

I propose, therefore, to regard events as occupying regions of space-time which, in some possible metric, are spheres so far as their space-dimensions are concerned, and between a certain maximum and a certain minimum so far as their time-dimension is concerned. The region "occupied" by an event is the class of points of which it is a member.

As the fundamental relation in the construction of points, we take a five-term relation of "co-punctuality," which holds between five events when there is a region common to all of them. A group of five or more events is called "co-punctual" when every quintet chosen out of the group has the relation of co-punctuality.

A "point" is a co-punctual group which cannot be enlarged without ceasing to be co-punctual.

In order to demonstrate the existence of points so defined, it is sufficient to assume that all events (or at least all events co-punctual with a given co-punctual quintet) can be well ordered. If Zermelo's axiom is true, this must be the case; if not, it may involve some limitation as to the number of events. I have been led by the arguments, first of Dr H. M. Sheffer, and then of Mr F. P. Ramsey, to the view that Zermelo's axiom is true; I am therefore less reluctant than I should have been formerly to assume that events can be well ordered.

To prove that every event is a member of at least one point, we proceed as follows—assuming that there are co-punctual quintets.

Let P be a well-ordered series whose field consists of all events; put

$$P = (x_1, x_2, \ldots x_n, \ldots x_\omega, x_{\omega+1}, \ldots x_\nu, \ldots)$$

Let a, b, c, d, y_1 be a co-punctual quintet. If y_1 is the only event co-punctual with a, b, c, d, then the class whose only members are a, b, c, d, y_1 is a point according to the definition. If, on the other hand, there are x's other than y_1 which are co-punctual with a, b, c, d, y_1, let y_2 be the first of them. If no x other than y_1 and y_2 is co-punctual with a, b, c, d, y_1 and y_2, then a, b, c, d, y_1 and y_2 form a point. Otherwise, let y_3 be the first x other than y_1 and y_2 and co-punctual with a, b, c, d, y_1, y_2; then y_3 must be later in the P-series than y_2. If this process comes to an end with y_n, then a, b, c, d, y_1, y_2, $\ldots$ y_n together form a point. If it does not come to an end with any finite n, it may happen that no x outside the series $(y_1, y_2, \ldots y_n, \ldots)$ is co-punctual with a, b, c, d and all the y's; in that case, a, b, c, d and these y's form a point. But if there are x's other than the y's and co-punctual with all of them, let y_ω be the first of them. Then y_ω is later in the P-series than any of the finite y's. We proceed in this way as long as possible, using two principles: (1) given a series of y's ending with y_ν, let $y_{\nu+1}$ be the first x in the P-series after y_ν and co-punctual with the group of all the previous y's; (2) given a series of y's having no last term, take as the next y the first x in the P-series which is after all the y's hitherto selected and co-punctual with all of them. If, at any stage, there is no such x, the y's already selected form a point. Now this process must end sooner or later; for the y's (other than y_1) form an ascending series selected from P, and therefore, sooner or later, there will be no x's later than all the y's previously selected. At this stage, if not before, a, b, c, d and the y's already selected will form a point. Hence if all events can be well ordered, every event is a member of at least one point, provided every event is a member of a co-punctual quintet. The proof still holds if we only assume that all events co-punctual with a given quintet can be well ordered.

Given any class of events α, let $R(\alpha)$ be the class of those events which are co-punctual with α. Then by definition α is a point if $\alpha = R(\alpha)$. The necessary and sufficient condition that all the members of α should have a point in common is that α should be contained in $R(\alpha)$. This condition is necessary, for, if δ is a point and α is contained in δ, it follows that $R(\delta)$ is contained in $R(\alpha)$, and that $\delta = R(\delta)$, so that α is contained in $R(\alpha)$. The proof that the condition is sufficient is longer; it is as follows.

If $\alpha = R(\alpha)$, α is a point. If not, let $S(\alpha)$ denote the part of $R(\alpha)$ which is outside α. Using again the P-series of all events, put

$z_1 =$ the first member of $S(\alpha)$ in the P-order.
$\zeta_1 = \alpha$ together with z_1.
$z_2 =$ the first member of $S(\zeta_1)$ in the P-order.
$\zeta_2 = \zeta_1$ together with z_2.
$\zeta_\omega = \alpha$ together with all the finite z's.
$z_\omega =$ the first member of $S(\zeta_\omega)$ in the P-order,

and so on, as long as possible. If $\mu < \nu$, z_μ precedes z_ν in the P-order. Hence, as before, there must come a stage when no fresh z's can be constructed. If ζ is the class consisting of α together with all the z's yielded by the method, ζ is a point. For (1) all the quintets in ζ are co-punctual, by the construction; (2) a term co-punctual with all the quartets of ζ cannot be later than all the ζ's, because if there were such a term we could construct more z's; (3) such a term cannot be earlier than some member of ζ because, if it were, it would have been chosen as the z of that stage in the construction; hence no event outside ζ is co-punctual with every quartet of ζ. Hence ζ is a point.

To say that a collection of events have a point in common is to say that the collection is part (or the whole) of the class which is the point. Conversely, a collection of events may contain a sub-class which is a point; the necessary and sufficient condition for this is that $R(\alpha)$ should be contained in α, where α is the collection in question. The proof proceeds exactly as before, if we now make $S(\alpha)$ mean the part of α which is not contained in $R(\alpha)$.

A group of events α is "co-punctual" if α is contained in $R(\alpha)$, and a "point" is a co-punctual group which cannot be enlarged without ceasing to be co-punctual.

A few purely logical properties of points may be noted. Given any two classes α and β, if α is contained in β, then $R(\beta)$ is contained in $R(\alpha)$. Hence if α and β are points and α is contained in β, α and β are identical; for in that case $R(\beta)$ and $R(\alpha)$ are respectively identical with β and α, and therefore if α is contained in β, β is contained in α, so that α and β are identical.

Every co-punctual group of events contains at least one point. This has already been proved, since to say that α is a co-punctual group is to say that α is contained in $R(\alpha)$.

It may be taken that, in general, there are a number of points of which any given event is a member. Such a set of points will fill a "region," but not every region will be the set of points to which some one event belongs. This topic, however, cannot be dealt with until we have discussed space-time order.

CHAPTER XXIX

SPACE-TIME ORDER

IN the present chapter I shall show how to develop spatio-temporal order, in the sense in which it is assumed by the general theory of relativity, without any apparatus beyond that of the preceding chapter, except a few hypotheses of the sort to be expected in founding *analysis situs*.

The transformations of co-ordinates which are admissible in tensor analysis are not unlimited; they are such, only, as leave relations of *neighbourhood* unchanged.* That is to say, a small displacement in one system of co-ordinates must correspond to a small displacement in any other. This requires that, independently of metrical considerations, the events of the space-time manifold should have certain relations of order. It must be possible, in certain circumstances, to say that A is nearer to B than to C, without presupposing any quantitative measure of distance. It must be possible to construct lines along which there is a definite order, but it must be impossible to distinguish certain lines as "straight." A closed curve will be distinguishable from an open curve, but two open curves will not be distinguishable from each other, provided they have no singularities. Generally, we shall be able to make propositions belonging to *analysis situs*, at any rate in a sufficiently small region. But propositions about a configuration must, in the geometry we are to construct, be only such as would remain true if the configuration were subjected to any kind of deformation which does not violate continuity. It is this pre-co-ordinate geometry that concerns us in the present chapter.

* For a geometry based on "neighbourhood," see Hausdorff, *Grundzüge der Mengenlehre* (Leipzig, 1914), chaps. vii. and viii.

The order to be introduced is of two sorts, macroscopic and microscopic. We will treat first of the former.

Let us observe, to begin with, that events may be divided into zones with respect to a given event. There are first those that are compresent with a given event, then those not compresent with it, but compresent with an event compresent with it, and so on. The nth zone will consist of events that can be reached in n steps, but not in $n-1$, a "step" being taken as the passage from an event to another which is compresent with it. We will call two points "connected" when there is an event which is a member of both. The passage from event to event by the relation of compresence may be replaced by the passage from point to point by the relation of connection. Thus points also can be collected into zones. If there is a minimum to the size of events, we may assume that it is always possible to pass from one event to another by a finite number of "steps." If so, there must be a smallest number of steps in which the passage can be made; thus every event will belong to some definite zone with respect to a given event. This is useful in the introduction of order, because we can agree that the mth zone is to be nearer the origin than the nth if $m<n$, so that it only remains to introduce order among the members of a given zone. And even here we only want such order as is involved in *analysis situs*, not such more rigid order as is involved, *e.g.*, in projective geometry.

When an event can be reached from another in n steps but not in $n-1$, we may regard the intermediate events as forming a sort of quantized geodesic route between the two events.

In virtue of the above division into zones, which can be effected with respect to any point as origin, we can define a rather small region of space-time by means of four integers, representing the number of steps in which any point in the region can be reached from four given points. It is only within a small region of this sort, therefore, that we need the more delicate methods of microscopic order, to which we shall now proceed.

Given two points κ and λ, let us denote by "$\kappa\lambda$" their logical product, *i.e.* the events which are members of both, or, in geometrical language, the events which contain both. It is obvious that, taking the view of events explained at the beginning of the preceding chapter, $\kappa\lambda$ will be null unless κ and λ are fairly near together. As already stated, we say that κ and λ are "connected" when $\kappa\lambda$ is not null. Microscopic order is confined to connected points, at any rate to begin with.

We now define "λ is between κ and μ" as meaning: "κ, λ, μ are points such that $\kappa\mu$ is not null and is a proper part of $\kappa\lambda$." An equivalent definition is: "κ, λ, μ are points such that $\kappa\mu$ is not null, and is contained in λ, but $\kappa\lambda$ is not contained in μ." By the help of suitable axioms, "between," so defined, can be made to give rise to the spatio-temporal order presupposed in assigning co-ordinates in the general theory of relativity. What the definition says, in geometrical language, is that every event which contains both κ and μ contains λ, but not every event which contains both κ and λ contains μ.

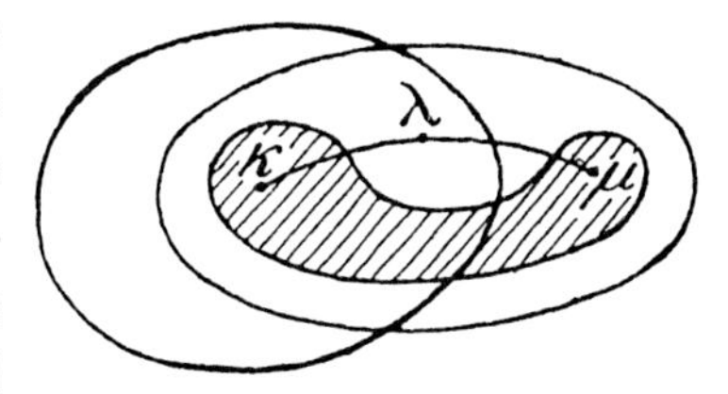

We must not imagine that all the points between two others lie on one line; each lies on *some* short route joining the end-points, a "short" route being one composed wholly of points between the end-points; but none lies on *all* short routes.

Before developing the formal consequences of this definition, it may be as well to consider its geometrical import. In the accompanying figure, λ will be between κ and μ if there are events which contain all three, but there are none which contain κ and μ without containing λ. (I represent events by areas.) Now if events can often be of irregular shapes such as that of the shaded area in the figure, it would seem that one event is not likely ever to be between two others according to the definition. I shall therefore assume that we may picture events as free from re-entrant angles and similar oddities. I imagine them as all oval; but formally it would do just as well if they were all four-dimensional cubes, and it would not matter whether they were large or small, provided they did not differ too much, and were all above a certain minimum. These pictorial requisites are rather for the *importance* of the theory to be developed than for its truth. In the preceding chapter, we assumed that events are such as to be all spheres according to one possible metric. Formally, we might equally well have assumed that there is a metric in which they are all cubes. Some assumption of this kind, as we saw, is necessary for the success of our definition of points. The other assumptions needed for its truth will be explicitly stated as they are introduced. The assumptions introduced so far in this chapter and its predecessor are:

(1) Compresence is symmetrical.

(2) Defining "events" as the field of compresence, every event is compresent with itself.

(3) Events can be well ordered; or at least those compresent with a given event can be.

(4) Any two events have a relation which is a finite power of compresence. (This is required for mapping space-time into zones.) In other words, the ancestral relation derived from compresence is connected.

We will now define a set of points as "collinear" if every pair of the set are connected, and every triad α, β, γ are such that either $\alpha\beta$ is contained in γ, or $\alpha\gamma$ is contained in β. We will define a set of points as a "line" if (1) it is collinear, (2) it is not contained in any larger collinear group with the same extremities. It will be seen that this definition is analogous to that of points. We may define a set of events as "co-punctual" when every quintet of the set are co-punctual; and we can then define a set of events as a "point" when (1) it is co-punctual, (2) it is not contained in any larger co-punctual group. This way of stating our previous definition of "points" brings out the analogy.

The "lines" that we are defining are not to be supposed "straight"; straightness is a notion wholly foreign to the geometry we are developing. Perhaps it might be better to call them "routes"; but there is no harm in calling them "lines" provided we remember that they are not supposed to be straight. For the present, we shall not be concerned with lines, but only with collinear groups of points.

Let us define a set of points as "α-collinear" if (1) every pair of the set is connected; (2) given any two, ξ, η, either ξ is between α or η, or η is between α and ξ. We shall want such axioms as will enable us to show that such a set of points is collinear, not merely α-collinear, and that their order is independent of α. It is obvious that, if we put ξ before η whenever ξ is between α and η, we obtain a serial order of any set of points which is α-collinear. But to insure that the order shall be independent of α we require the following three axioms:

(1) If α, β, ξ, η are points, and $\alpha\beta$ is contained in $\xi\eta$, and $\alpha\eta$ is contained in ξ, and ξ and η are distinct, then $\beta\eta$ is not contained in ξ.

(2) If $\alpha\eta$ is contained in ξ, and $\beta\xi$ is contained in η, than $\alpha\beta$ is contained in the sum of ξ and η. (It follows at once that $\alpha\beta$ is contained in $\xi\eta$.)

(3) If $\alpha\beta$ is contained in η, and $\alpha\eta$ is contained in ξ, then $\beta\xi$ is contained in the sum of α and η. (It follows at once that $\beta\xi$ is contained in η.)

The practical effects of these three axioms are:

(1) If ξ and η are between α and β, and ξ is between α and η, then ξ is not between β and η.

(2) If ξ is between α and η, and η is between β and ξ, then ξ and η are between α and β.

(3) If η is between α and β, and ξ is between α and η, then η is between β and ξ.

From these axioms we can deduce that a set of points which is α-collinear is collinear. Also that, given a set of α-collinear points, if γ is one of them, the points of the set which are beyond γ from α are γ-collinear, and retain the same order when arranged with reference to γ as they had when arranged with reference to α. Also that, if β is one of a set of α-collinear points, those of the set which are between α and β are β-collinear, and have, when arranged with reference to β, the converse order to that which they had when arranged with reference to α. These propositions show that we have a satisfactory definition of order among the points of a collinear set.

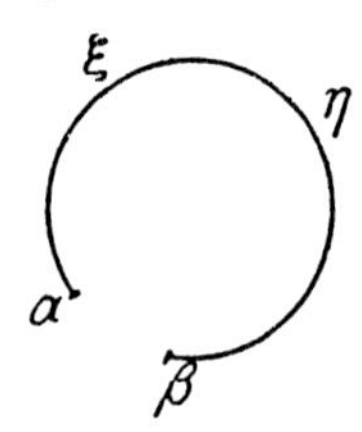

The above axioms are logically adequate, but regarded as asserting physical truths about events they may perhaps be regarded as more or less doubtful. We have to remember that our lines are not straight, and may therefore return into themselves. Routes with very great curvature are, however, excluded by our definition of collinearity. Consider, *e.g.*, such a route as that in the accompanying figure. We may suppose that α, β, ξ, η are all connected, but ξ and η will not be between α and β according to the definition, because obviously an event may contain α and β without containing ξ and η. Thus if we wish to regard the above route from α to β as, in *some* sense, a line, it will have to be in an extended sense, namely, that it can be divided into a number of small finite parts, each of which is a line. And a set of points may be regarded as collinear in an extended sense if it is capable of a serial order such that any sufficiently small consecutive stretch of the series is a collinear set—provided that such stretch must contain not less than four points.

We can now prove, by the help of one further axiom, that any progression of collinear points all lying between two points α and β must have a limit.

Let our set of points be $\kappa = (\xi_1, \xi_2, \xi_3, \ldots \xi_n \ldots)$, all lying on a line between α and β, in an order from α towards β. Let σ be the sum of all the points in κ (*i.e.* the class of members of members of κ), and $\tilde{\omega}$ their product, *i.e.* the events which belong to every member of κ. Then $\tilde{\omega}$ is not null, because $\alpha\beta$ is contained in it, and α, β are connected (in virtue of the definition of collinearity).

Let κ_1 consist of all the ξ's except ξ_1, κ_2 of all the κ_1's except ξ_2, etc. Let $\tilde{\omega}_1$ be the events belonging to all members of κ_1, and generally let $\tilde{\omega}_n$ be the events belonging to all members of κ_n; and let λ be the sum of all the $\tilde{\omega}$'s. Then λ consists of all those events which belong to all sufficiently late ξ's; *i.e.* to say that an event is a member of λ is to say that there is an n such that the event is a member of $\xi_{n\ m}$ for all values of m.

It will be observed that κ_{n+m} is contained in κ_n, therefore $\tilde{\omega}_n$ is contained in $\tilde{\omega}_{n+m}$. It follows that, if z, z' are two members of λ, there is an n such that z, z' are both members of $\tilde{\omega}_n$. Hence they are both members of ξ_{n+1}. Hence any five members of λ are co-punctual, and therefore there is at least one point which contains the whole of λ, since λ is contained in $R(\lambda)$.

If there is a limit, say δ, to the series of ξ's, we require:

(1) That δ should be beyond all the ξ's, *i.e.* that for every n and m we should have $\delta\xi_n$ contained in ξ_{n+m}, *i.e.* that we should have $\delta\sigma$ contained in λ;

(2) That there should be no point beyond all the ξ's but between them and δ, *i.e.* that, if η is any point such that $\eta\sigma$ is contained in λ, then $\eta\sigma$ is contained in δ.

A sufficient condition is, therefore, $\delta\sigma = \lambda$. If there is a point δ fulfilling this condition, it is the required limit.

If there is an event z such that every quartet of λ is co-punctual with z and every quartet of σ which is co-punctual with z is a part of λ, then there is a point δ which contains λ and has z for a member, and this point will be such that $\delta\sigma = \lambda$, so that it will be the required limit. But if there is no such event as z, we must proceed differently.

In this case we need a new axiom, namely:

If β is between α and γ, and x is a member of α but not of β, then there is a quartet which is contained in β and γ but is not co-punctual with x.

In the figure, y represents a member of such a quartet.

Given this axiom, we proceed as follows.

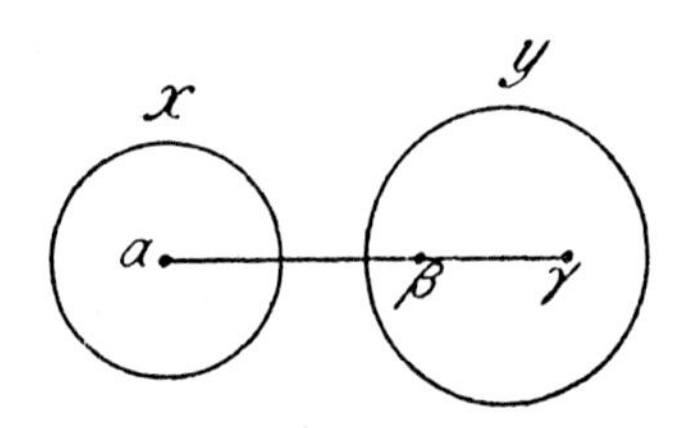

Since ξ_{n+1} is between ξ_n and β, if x is a member of ξ_n but not of ξ_{n+1}, there is a quartet which is contained in β and ξ_{n+1}, but is not co-punctual with x. Now $\beta\xi_{n+1}$ is contained in λ; therefore there is a quartet which is a part of λ but is not co-punctual with x. It follows by transposition that if x is a member of ξ_n and every quartet of λ is co-punctual with x, then x is a member of ξ_{n+1}. It follows that x is a member of $\xi_{n+2}, \xi_{n+3}, \ldots$ so that x is a member of λ. Hence, since ξ_n may be any member of κ, it follows that any member of σ which is co-punctual with the whole of λ is a member of λ. Now the terms co-punctual with the whole of λ constitute the class $R(\lambda)$. Hence the common part of σ and $R(\lambda)$ is contained in λ, and is therefore equal to λ, since λ is contained in σ and in $R(\lambda)$.

Now if δ is a point which contains λ, it follows that δ is contained in $R(\lambda)$; hence $\delta\sigma$ is contained in λ, and is therefore equal to λ, since λ is contained in δ and in σ. Hence δ is the required limit.

It follows from this that a compact series of points contained within a stretch of collinear points is continuous. It does not follow that there are compact series of points; this would require existence-axioms which there is no object in introducing, since we do not know whether space-time is continuous or not. It is, however, interesting to observe that an initial apparatus of $\aleph_0$ events suffices to generate a continuous space-time of points, by means of the relations of co-punctuality and logical inclusion.

The further development of our geometry, so as to include surfaces, volumes, and four-dimensional regions, obviously presents no difficulty in principle, and I do not propose to enlarge upon it. I will merely observe that it is possible to extend the method by which we have defined points and lines so as to obtain something which we may call surfaces and regions, though not quite in the usual sense. Probably various ways of doing this are possible; the one that I suggest is the following.

A class of lines will be called "co-superficial" when any two intersect, but there is no point common to all the lines of the class.

A "surface" is a co-superficial class of lines which cannot be augmented without ceasing to be co-superficial.

A class of surfaces is "co-regional" when any two have a line in common, but no line is common to all the surfaces of the class.

A "region" is a co-regional class of surfaces which cannot be augmented without ceasing to be co-regional.

It is obvious that this method could be extended to any number of dimensions; also that it requires limitations and extensions. But it seems unnecessary to pursue the matter further, since it is plain that we have what is needed for the pre-co-ordinate geometry of space-time.

Let us now compare our constructed space-time with the spatial manifolds of *analysis situs*. In the preceding chapter we quoted Hausdorff's definition of a "topological" space, and we saw that, in order to prove the usual propositions about limits, it is necessary that the total number of neighbourhoods should be $\aleph_0$. Let us now define as a "neighbourhood" of a point x any set of points each of which contains as a sub-class a certain finite co-punctual class of events which is a sub-class of x. That is to say, if α is a co-punctual class of events each of which is a member of x, the set of all the points of which α is a sub-class will be a neighbourhood of x. With this definition of a "neighbourhood," it is obvious that our space has the four characteristics by which Hausdorff (*loc. cit.*, p. 213) defines a topological space. In order to insure that our space shall also satisfy his second denumerative axiom (*loc.cit.*, p. 263), it is necessary and sufficient to assume that the total number of events is $\aleph_0$. With this assumption, the theorems of *analysis situs* become applicable to our space-time manifold of points.

It remains to say a word on the subject of dimensions. We have not so far said anything explicit on this subject, though our original introduction of co-punctuality as a five-term relation could only prove satisfactory in a four-dimensional manifold. The most suitable definition of dimensions from our point of view is that of Poincaré, which is inductive. He defines a space M as one-dimensional if, given any two points P, Q, there is an isolated set of points X such that no connected part of M-not-X contains both P and Q. And he defines a space M as n-dimensional if, given any two points P, Q, there is an $(n-1)$-dimensional set of points X such that no connected part of M-not-X contains both P and Q. Using this definition, or any other which is purely topological, we set up the axiom that our topological space-time is to be four-dimensional.* This completes the material required for the topological treatment of space-time.

* For an account of the modern theory of dimensions, see Karl Menger, *Bericht über die Dimensionstheorie*, Jahresbericht der deutschen Mathematiker-Vereinigung, 35, pp. 113-150 (1926).

CHAPTER XXX

CAUSAL LINES

THE notion of causality has been greatly modified by the substitution of space-time for space and time. We may define causality in its broadest sense as embracing all laws which connect events at different times, or, to adapt our phraseology to modern needs, events the intervals between which are time-like. Now owing to the fact that the formula for ds^2 is formally the same for time-like and for space-like intervals, there is no longer the difference that formerly existed between causal and geometrical relations. Geodesics are geometrical, but they are also the paths of material particles. It is hardly correct to say that a particle *moves* in a geodesic; it is more correct to say that a particle *is* a geodesic (though not all geodesicis are particles). To say that a particle moves in a geodesic is to use language appropriate to the conception of a space which persists through time, involving the notion of a position which may be occupied either at one time or at another. We think, for example, that it is possible to move from A to B or from B to A; but such a view is incompatible with the theory of space-time. According to that theory, every position of a body has a date, and it is impossible to occupy the same position at another date, since the date is one of the co-ordinates of the position. When we travel from A to B, the date is continually advancing; the return journey, having different dates, does not cover the same route. Thus geometry and causation become inextricably intertwined.

Dr A. A. Robb has laid stress upon the fact that, when two events have a space-like interval, there can be no direct causal relation between them. This means that, given two such events A and B, if any inference is possible from the one to the other, it must be by way of a common causal ancestor. Two men may see the sun at the same moment, so that the interval between their percepts is space-like; the inference that so-and-so is seeing the sun now arises from our knowledge of radiation, and requires that we should trace his percept and our own to a common ancestry in the sun. We may therefore distinguish time-like and space-like intervals by saying that the former occur where there is some direct causal relation, while the latter occur where both events are related to a common ancestor or a common descendant. And possibly the magnitude of the interval may be derivable from the magnitude of the causal relation. But if this is to be possible, it will be necessary to achieve considerable precision as to what we mean by causal relations.

As we saw in Part II., perception as a source of knowledge concerning physical objects would be impossible if there were not, in the physical world, semi-independent causal chains, or causal lines as we may call them. The light which comes to us from a printed page retains the structure of the page; if it did not, reading would be impossible. The retention is only approximate; it ceases at a distance from the book. And it ceases within the eye if we have defective vision. But where there is such failure, perception ceases—or rather, it fades away as the failure to preserve structure increases. Thus it is essential to perception as a source of knowledge that there should be in the world causal series which are, within limits, independent of the rest of the world.

Another point concerning causation emerges from the consideration of perception. A number of simultaneous percepts—*e.g.* the letters of a word which we read at a glance—are to be regarded as "co-punctual" in the sense of our two preceding chapters. Each of these percepts has its own causal antecedents, different from those of the other percepts. It is true that there may be mutual modification—*e.g.* a colour looks different in the neighbourhood of another colour from what it looks against a dark background. But this is recognized as "modification," *i.e.* as effecting a change from a norm, which must remain within limits if perception is to be successful. Thus the percipient is the meeting-place of a number of more or less independent causal series—as many, at least, as there are distinguishable elements in his total momentary perceptual field. But although these lines have converged upon him more or less independently, the totality of his percepts now becomes a causal unit, as is seen in mnemic phenomena. Given a number of simultaneous percepts, a percept very similar to one of them, occurring on a future occasion, recalls something similar to the others, or at least may do so; here the co-punctuality of the percepts is essential to the character of their total effect.

In the physical world, the same sort of thing must be supposed to occur, though to a less striking degree. According to the theory of Chapter XXVIII., any event in the physical world occupies a finite region of space-time, whose finiteness consists in the fact that the said event is compresent with events which are not compresent with each other. On the analogy of mnemic phenomena, a group of co-punctual events may have effects which would have been impossible if the events had not been co-punctual. That is the reason why physics is compelled to resort to points in stating its causal laws. Until we have a complete group of co-punctual events, *i.e.* a point, we cannot be quite sure as to the effect which will follow from any one of the events; such knowledge as we can have will be more or less approximate.

It is these two opposite laws, of approximately separable causal lines on the one hand, and interactions of co-punctual events on the other, which make the warp and woof of the world, both physical and mental. In this chapter, I want to attain more precision as to the separable causal lines.

The possibility of perception, as we have seen already, depends upon the occurrence in the physical world of processes which may be called "radiations," provided the word is used somewhat more widely than is customary. The processes commonly called radiations are, naturally, the most perfect examples. In these, when they are undisturbed, we have a condition of some kind which spreads outward from a centre, changing in an apparently continuous manner as it travels. Something may be met with on the way which alters the law of change, or even stops the radiation in some direction altogether; but in the absence of obstacles the process proceeds according to its own intrinsic laws. The public senses—sight, hearing, and smell—depend upon radiations, in a generalized sense in the case of smell. Bodily senses, including touch, are more analogous to electric currents in their manner of propagation: they travel along nerves, but not through air or empty space. The public senses, also, travel along nerves, but the disturbance in the nerves is a prolongation, with alterations, of a process in the world outside the percipient's body, which is not the case with the bodily senses. It is owing to the existence of radiations that we live in a common world, since this depends upon the fact that neighbouring percipients receive similar stimuli at about the same time. The physical account of radiations is, however, very different in different cases. In the case of smell, the emission theory is universally accepted: we smell a body because portions of it travel from it to the nose. In the case of sound, only a process, not actual matter, is transmitted, but the process is in matter. In the case of light, if we accept the undulatory theory, the process consists of a transverse vibration, which may be said to be in the æther if that brings comfort to the speaker, but is certainly not in ordinary matter. If we could accept the light-quantum theory, we should still suppose that there is some periodic process, such that the action during one period is *h* (Planck's constant); the light consists of (so to speak) atoms, each of which is such a process. There is a great difference of physical importance between these three cases of smell, sound, and light; the first is quite unimportant physically, the second a somewhat late development from more fundamental principles, the third a corner-stone of physical theory.

In the ideal case of a radiation, a few observations should suffice to determine its centre, and then, its laws being known, we could infer the whole connected system of events which constitutes it, in so far as the events enter into physical laws. The case of light from a fixed star very nearly realizes the ideal. The places in the universe where the light encounters obstacles are very few, though unfortunately they include the places where we live. It is because this example of light *in vacuo* is so nearly perfect that we know as much as we do about astronomy.

Radiation independent of matter, however, is only one form of causal process in the physical world. Apart from quantum changes, there are at least two others which are of great importance: one is the motion of matter, and the other is the transmission of a process by matter. The difference involved is essentially one as to causal laws: one sort of causal connection between events makes us regard them as part of the history of one piece of matter, while another does not, but there is no more intimate connection between an electron at one time and the same electron at another time than between two parts of one light-ray. Let us consider for a moment the nature of the causal laws which define one piece of matter.

One *prima facie* difference is that the propagation of light is spherical (or conical, in the case of a directed beam), whereas the motion of matter is linear. The history of a piece of matter is a "world-line"; the history of a light-wave is not. This difference may no longer exist if some adaptation of the light-quantum theory can be made satisfactory; but, if so, we shall feel that the difference between light and matter has been much diminished. Another difference is the relative indestructibility of matter. One form of energy changes into another, but the energy represented by the proper mass of an electron or proton is not *known* to change into other forms, and apparently never does so under terrestrial conditions: it does not radiate at all in any circumstances that we can produce or observe. Then there is the fact that the velocity of a body relative to any observer is always less than that of light. But in spite of the doubt as to light-quanta, the main feature of the causal laws that constitute matter seems to be their linear rather than spherical character. It is this that enables us to locate a given piece of matter at a given time. The light emitted by a flash is, at a given moment, diffused over the surface of a sphere, but an electron is as concentrated at one time as at another, and does not tend to spread itself out. A unit of matter may, therefore, be appropriately defined as a "causal line."

Before pursuing this subject, however, it will be well to dispose of the other kind of causal process which we mentioned just now, namely the transmission of a process by matter. This is itself of two sorts, one illustrated by sound, the other by the conduction of an electric current. In the case of sound we have a radiation; in the other case we have a more or less linear process. In each case, however, actual pieces of matter move, and cause others to move. The former belongs to the notion of a "causal line," to which we shall return in a moment. The latter belongs to the causal laws as to the interactions of different pieces of matter, which I do not wish to consider until I have elicited the intrinsic causal laws which constitute the definition of one piece of matter. These, as we saw, have been somewhat obscured by the notion of substance, which made it plausible to take for granted certain connections between events at different times, which, for us, are causal, and demand explicit recognition. It is these intrinsic laws which replace substance that I wish to consider now, leaving the interactions between different pieces of matter for a later stage.

What, then, constitutes a "causal line"? In other words, what constitutes one electron? Before asking ourselves what makes us call an electron at one time the same as an electron at another time, it may be well to ask ourselves: What constitutes an electron at one time?

We must find some reality for the electron, or else the physical world will run through our fingers like a jelly-fish. There is the same sort of reason, however, for not regarding an electron as an ultimate particular as there was for refusing this status to a space-time point. The electron has very convenient properties, and is therefore probably a logical structure upon which we concentrate attention just because of these properties. A rather haphazard set of particulars may be capable of being collected into groups each of which has very agreeable smooth mathematical properties; but we have no right to suppose Nature so kind to the mathematician as to have created particulars with just such properties as he would wish to find. We have, therefore, to ask ourselves: Can we construct an electron out of events, in the same sort of way in which we constructed space-time points? To this inquiry we must now address ourselves, confining ourselves, at first, to the electron at one time.

When I speak of "electrons" in this discussion, I shall include "protons," since everything that is to be said about the one is to be said about the other also.

We do not know much about the contents of any part of the world except our own heads; our knowledge of other regions, as we have seen, is wholly abstract. But we know our percepts, thoughts, and feelings in a more intimate fashion.

Whoever accepts the causal theory of perception is compelled to conclude that percepts are in our heads, for they come at the end of a causal chain of physical events leading, spatially, from the object to the brain of the percipient. We cannot suppose that, at the end of this process, the last effect suddenly jumps back to the starting-point, like a stretched rope when it snaps. And with the theory of space-time as a structure of events, which we developed in the last two chapters, there is no sort of reason for not regarding a percept as being in the head of the percipient. I shall therefore assume that this is the case, when we are speaking of physical, not sensible, location.

It follows from this that what the physiologist sees when he examines a brain is in the physiologist, not in the brain he is examining. What is in the brain by the time the physiologist examines it if it is dead, I do not profess to know; but while its owner was alive, part, at least, of the contents of his brain consisted of his percepts, thoughts, and feelings. Since his brain also consisted of electrons, we are compelled to conclude that an electron is a grouping of events, and that, if the electron is in a human brain, some of the events composing it are likely to be some of the "mental states" of the man to whom the brain belongs. Or, at any rate, they are likely to be parts of such "mental states"—for it must not be assumed that part of a mental state must be a mental state. I do not wish to discuss what is meant by a "mental state"; the main point for us is that the term must include percepts. Thus a percept is an event or a group of events, each of which belongs to one or more of the groups constituting the electrons in the brain. This, I think, is the most concrete statement that can be made about electrons; everything else that can be said is more or less abstract and mathematical.

We have arrived at the conclusion that an electron at an instant is a grouping of events; the question is: what sort of group is it? Obviously it includes all the events that happen where the electron is. If we may regard the electron as a material point, the events constituting an electron will have the two characteristic properties of points, viz. any five are co-punctual, and not all sub-classes of four events are co-punctual with any event outside the group. I do not know whether there is any valid ground for supposing that an electron is of finite size; none of the usual arguments seem at all conclusive, since they only show the forces developed in the neighbourhood of an electron. However, it is usual to assume a finite size, and for us the matter is one of indifference. If we assume a finite size, the events belonging to the electron can be grouped into many points, not only into one; in this case, the electron is a group of points, *i.e.* a class of classes of events. It will save circumlocution to speak of the electron as a point, and leave it to the reader to make the necessary verbal alterations for adaptation to the hypothesis of finite size. But it should be remembered that in Heisenberg's theory the electron is neither a point nor of finite size, since ordinary spatial conceptions are inapplicable to it. For the moment, we will, however, confine ourselves to the older theory of the electron.

If the electron is a point, it is a *material* point, and thus differs from points in empty space. This difference, I believe, does not consist in anything characteristic of the electron at an instant, but in its causal laws. What distinguishes a material point from a point of empty space-time is that we can recognize a series of earlier and later material points as all parts of the history of one electron. In the Newtonian theory, one could say the same of a point of absolute space; but with the abandonment of absolute space we have become unable to regard a point at one time as in any sense the same as a point at another time, except in the case of a *material* point. The existence of this connection may be taken as the definition of "matter"; and obviously the connection is causal.

In order to develop this further, we must return to the view suggested in connection with perception, that events occur, usually, in groups arranged about centres. These centres may be taken to be places where there is matter. It is found that, given events arranged about a centre at one time, there are generally similar events arranged about neighbouring centres at slightly earlier or later times. By taking the centre very small, and by continually diminishing the time-like interval concerned, this statement can be made more and more nearly true; in the limit, when stated in the language of differentials, it *may* be exactly true, except where quantum phenomena are concerned. In their case, continuity is not the criterion, at least not continuity in all respects. There is continuity in some respects, and in others there is a jump of a definite amount connected with the quantum theory. This case shows, however, that continuity is not the essence of material identity; the essence is inferribility of a group of phenomena at one time from a group at another, when both groups are arranged about centres.* The time must be very short, and the inference is only approximate, except in the limit, as the time tends towards zero. Moreover, the time of the group is not any of the times at which the several members of the group occur, but the calculated time at which the group began to be propagated from the centre. The centre is "where the piece of matter is," and the route of the piece of matter is determined by the differential equations which result from the above principle. But as to what are the actual events at the centre, we know nothing except what follows from the fact that our percepts and "mental states" are among the events which constitute the matter of our brains.

Thus each material unit is a causal line whose neighbouring

* In this case, however, if Heisenberg is right, we cannot identify an electron at one time with an electron at another. This would be a difficulty if an electron were conceived as a substance, but for us it is merely an empirical limitation of the empirical conception of a causal line.

points are connected by an intrinsic differential law. The simplest form of such a law is the first law of motion, from which it follows that if a body covers a given distance in a very short time, it will cover a very nearly equal distance in the next very short time. I conceive—though this is conjectural —that, given any event anywhere in space-time, there is usually some qualitatively very similar event in a neighbouring place in space-time, and that, if there is any measurable relation between the two events, the " velocity " of the change varies continuously, so that at a third neighbouring point there will be an event differing from the second by very nearly the same amount as that by which the second differed from the first, provided the interval between the second and third points is equal to that between the first and second. This, together with the fact that events can be grouped about centres by the sort of laws which we have called " perspective," seems to explain the utility of matter in stating the causal laws of the physical world. But there is need of caution owing to quantum phenomena, as explained in the preceding paragraph. Continuity is the rule, but it may have exceptions. So long as the exceptions are subject to ascertainable laws, they do not make the whole system impossible.

So far, I have said nothing about extrinsic causal laws, *i.e.* those which we naturally regard as exemplifying the influence of one piece of matter upon another. Einstein's theory of gravitation has thrown a new light upon these; but this is matter for a new chapter.

CHAPTER XXXI

EXTRINSIC CAUSAL LAWS

I MEAN by an "extrinsic" causal law any formula in which one piece of matter is mentioned as concerned in the behaviour of another. Newtonian gravitation afforded a perfect example of an extrinsic causal law, but Einsteinian gravitation, *prima facie*, does not. The question I want to consider is: Can we, in the last analysis, dispense with such laws altogether, and regard each piece of matter as completely self-determined? Or must we admit them, and, if so, in what form? And what are we to say of such matters as the emission and absorption of light?

Let us first consider Einsteinian gravitation. The theory consists in ascribing to every region of space-time a metrical structure which is obtained (roughly speaking) by superposing a number of structures which are symmetrical about centres, the centres being portions of matter; and, given the structure, each piece of matter moves in a geodesic, or rather is a geodesic. It is not very easy to see what this means when it is translated from the technical language of theoretical physics into the language of groups of events. Nevertheless, we must make the attempt.

To begin with: Can we make "matter" into a mere law according to which events occur in the places where there is no matter? This question is analogous to that of phenomenalism as discussed in Chapter XX. We there considered the possibility of explaining unperceived "things" as laws concerning the behaviour of perceived "things." Similarly we might take events which occur in empty space, and find that they were subject to laws symmetrical about centres, and define each such law as a piece of matter situated at the centre. Conversely, we might regard the supposed events in empty space as mere laws connecting events in different pieces of matter; this becomes phenomenalism if we confine the pieces of matter to human brains. There are many possible ways of turning some things hitherto regarded as "real" into mere laws concerning the other things. Obviously there must be a limit to this process, or else all the things in the world will merely be each other's washing. But the only obvious final limit is that set by phenomenalism—perhaps one ought to say, rather, that set by solipsism. If we have once admitted unperceived events, there is no very obvious reason for picking and choosing among the events which physics leads us to infer.

This argument, however, hardly warrants us in assuming events inside an electron. If we assume an electron of the Rutherford type, we shall have to say that, if anything does take place inside the electron, we can know nothing about it. No physical process passes through the electron, so that the inside, if it exists, is a prison from which nothing can escape. No event inside an electron can be compresent with an event outside it; consequently, according to the theory of Chapter XXIX., no line can cross the boundary of an electron. What goes on inside, if anything does, is irrelevant to the rest of the universe, and is not really in the same space-time as what goes on outside. Now the world of physics is intended to be a causally interconnected world, and must be such if it is not to be a groundless fairy tale, since our inferences depend upon causal laws. Therefore if anything occurs which is causally isolated, we cannot include it in physics. We have no ground whatever for saying that nothing is causally isolated, but we can never have ground for saying: Such-and-such a causally isolated event exists. The physical world is the world which is causally continuous with percepts, and what is not so continuous lies outside physics. Thus if anything occurs inside an electron, such an occurrence does not belong to the world of physics. It would seem to follow that, if the electron is to have a definite position in space-time, it must be either a point or a hole. The former, however, is physically unsatisfactory, and the latter seems scarcely capable of an intelligible interpretation. Thus the Rutherford type of electron raises problems, however we may interpret it.

The Heisenberg electron offers a way out of these difficulties. This electron is not in a definite place, and nothing happens inside it. It is essentially a collection of radiations observable in other places than that in which the electron would formerly have been said to be. Thus the electron is reduced to a law as to occurrences in a certain region. We cannot say, on this view, that the electron is a point, or that it is a certain finite region, or that it is a hole; it is, so to speak, something of a different logical type, connected with a region through the fact that the radiations concerned have diminishing intensity as we pass away from this region, but not capable of *accurate* correlation with either a region or a point. Thus on this view matter consists merely of laws as to occurrences in " empty " space.

Owing to the fact that an electron at one time cannot be identified with an electron at another time where quantum changes have intervened, the conception of motion loses its definiteness where electrons are concerned. This, however, only raises difficulties when we are concerned with very minute phenomena, such as those which occur within an atom. For large-scale phenomena, such as those with which astronomy is concerned, we may still regard the electron as persisting and as moving in space-time.

We can now return to the Einsteinian theory of gravitation, which necessitated this long digression. According to this theory, each electron is associated with a crinkle, which grows less marked as we get away from the electron, but extends theoretically throughout space. The actual metrical structure of space-time in any region is obtained (roughly speaking) by superposing these crinkles. Now the metrical properties of space-time are nothing but a method of stating causal laws. In the case of gravitation, these laws have to do with the way in which the movement of one electron is connected with the positions of the others. We must suppose that the formula for interval represents something in the state of affairs at each place, and that bodies left to themselves move in geodesics, and that, so long as electromagnetic phenomena are left out of account, the formula for interval at any place is found approximately by superposing a number of spherically symmetrical formulæ, each of which corresponds to an electron in its central region. It is natural to ask, at this point, whether interval has any more physical reality than force. But I do not wish to raise this question yet, as I propose to consider it in later chapters. For the present we may say (*a*) that we can recognize peculiar regions in space-time, which are those that would naturally be regarded as in the immediate neighbourhood of matter; (*b*) that the formula for interval at any place is a function of the geodesic distances from that place to neighbouring pieces of matter; (*c*) that pieces of matter travel along geodesics.

The question whether, in such a theory, there is "action at a distance" is really one of words. The formula by which we determine what will happen in a given region will contain references to distant regions, and it may be said that this is all we can mean by "action at a distance." To mean more, it may be said, is to regard causality as something more than correlation, which there can be no reason for doing. If what happens in one place is correlated with what happens in another, we may be told, nothing more could be imagined in the way of action at a distance. But this is not quite what in fact occurs. What happens in one place is not correlated with *what happens in* another place, but with another place, which is a different thing. Different neighbourhoods have different characters, and the differences can be represented by a combination of formulæ which are spherically symmetrical. This is not action *at* a distance, but action *according to* a distance; there is nothing that can properly be called an effect of one thing upon another at a distance from it. Thus so far, pending the discussion of interval, we have found nothing that can properly be described as an extrinsic causal law.

Electromagnetic phenomena, if we accept Weyl's theory, will not differ importantly, so far as our present question is concerned, from gravitation. An electromagnetic field will be represented by gauge-relations between points in a neighbourhood, and there will be no ground for supposing that one piece of matter influences another; all that we can say is that a piece of matter corresponds to a metrical state of affairs which makes the geodesics different from what they would otherwise be. The motion of an electron or proton is then due to the peculiarities of the metrical state of affairs where it is, not to something even so near as the hydrogen nucleus is to its planetary electron.

But what are we to say of the emission and absorption of light? It is clear that whenever we perceive light we absorb it, that is to say, the energy in the waves of light (or light-quanta ?) that hit the eye is transformed into a different kind of energy, though I should not venture to say what kind. Therefore all visual percepts involve this process of absorbing light. And if perception can ever be a source of knowledge as to things outside the percipient's body, there must be causal laws connecting what happens to the percipient with what goes on outside. It is, of course, obvious that there are such laws; we cannot revive Leibniz's windowless monads. The process of absorption and emission of light will serve as a special case, about which we have considerable knowledge, in which we can hope to analyze exactly what occurs.

Let us take, for simplicity, two hydrogen atoms, of which one emits energy which the other absorbs. But for the theory of quanta, and such phenomena as the photo-electric effect, a supposition of this sort would be impossible. If the energy radiated from a hydrogen atom in the form of light really has the shape of a spherical wave, it is impossible that the whole of it should be absorbed by one other atom, any more than the whole of the light radiated from the sun can fall on the earth. But if the light emitted by a single atom travels in a straight line (approximately), like a material particle, then it may happen to hit one atom and be absorbed whole, just as Jonah might have been swallowed by another whale. We shall have to suppose, in this case, that the spherical distribution of light round a radiating body is a statistical phenomenon, like bullets fired from a fort in all directions. This suggests the hypothesis which we have already considered in Chapter XIII., according to which nothing at all happens between the emission of light by one body and its absorption by another. In that case, empty space collapses just as the electron did, and only the surface of the electron remains. This, however, seems hardly a tenable view. The intervening space might be described as non-existent from a metrical point of view, since the interval between the emission and the absorption of a light-ray is zero; but from an ordinal point of view this is not the case, since, if A and B are two points on a light-ray, we can distinguish the case in which the ray goes from A to B from that in which it goes from B to A. This difference can be stated in metrical terms. For example: Let us take as our time co-ordinate the proper time of no matter what body; whatever body we choose, A will be earlier than B, or else, whatever body we choose, B will be earlier than A. Again: Suppose that at A and B there are mirrors, which reflect part of the ray in such a way that an observer O sees both reflected rays. Then either every such observer will see the reflection from A before that from B, or else every such observer will see the reflection from B before that from A. We can free this from dependence on an observer by the following method of statement: Let A' be a point on the ray reflected from A, and B' a point on the ray reflected from B, so chosen that the interval between A' and B' is time-like. Then, however A' and B' may be chosen, either A' is always before B', or B' is always before A'. This is stated in the language of the special theory, but it is still valid, *mutatis mutandis*, in the general theory. Thus when we say that the interval between two points on a light-ray is zero we are not denying that there is an important sense in which one is earlier than the other, and in which one can be regarded as cause and the other as effect. This suggests that the zero interval is not quite so significant as it might seem to be, and I cannot therefore accept the view that there are no events along the path of a light-ray in empty space.

Let us now return to the emission of light, ignoring absorption for the present; and let us still consider a single hydrogen atom. We are told to suppose that the electron revolves about the proton for a certain time, say in a circular orbit four times as large as the minimum orbit; then, suddenly, it decides to revolve in the minimum orbit. When this change occurs, the atom loses a certain amount of energy, which is transformed into light whose frequency is obtained by dividing the loss of energy by h (Planck's constant). Whether the light travels only in one direction, or in a spherical wave, we are compelled, in the present state of physical knowledge, to leave an open question. But we do assume that something travels away from the electron, and that, if light is absorbed by another atom, that light has traversed a route from its place or places of origin. We assume also that the light has a frequency, *i.e.* that what travels is a periodic process. When the light is absorbed, it ceases to exist as light, although it may reappear (in fluorescence). But often its energy exists in discoverable forms—chemical forms in chlorophyl, for example. When, however, the energy exists in the form of a steady motion of the electron in its orbit, it is not discoverable until there is a change of orbit. If we had sufficiently powerful microscopes, we could see a glowing gas dissolving into a comparatively small number of spots of light, while the atoms in steady motion would be invisible. Thus we seem to reach the conclusion that the causal laws which genuinely connect one piece of matter with another are quantum laws, in which there are various stages: first, a periodic process having no outside effect; secondly, a sudden disruption of the energy of this process into two parts, one being a new periodic process in the original body, the other a periodic process travelling in empty space; thirdly, the arrival of the travelling process at another body; fourthly, a quantum change in this other body, involving absorption of the radiant energy in the production of a new steady state in the absorbing body. All genuine causal relations between different bodies, we may suppose, involve this process of sudden loss of energy by one body and its sudden acquisition, later, by another body. The older physical laws, as reinterpreted by relativity, can apparently be so stated as to leave bodies independent of each other; but I cannot see how the quantum laws can be so stated.

If one could adopt what may be called the "parcels-post" theory of radiation, according to which, when energy leaves an atom, it does so with a definite destination in view, we could simplify our account of the matter. In that case, atoms would, at most times, live a self-contained life, "the world forgetting, by the world forgot." But sometimes they would give a parcel of energy to the postman, and sometimes they would receive one from him. The postman (who is perhaps not a teetotaller) sways from side to side as he travels, and the bigger the parcel the faster he sways. But he travels at the same rate whether his parcel is big or small; and he is the only link between the atom and the rest of the world.

For the present, we dare not assume that the question is as simple as in the parcels-post illustration. Energy may (as the orthodox theory supposes) be lost by radiation into the void—lost, I mean, not mathematically, but practically. The difficulty is that we cannot put an instrument into the void to see what happens there; the attempt is just like trying to go and see what things look like from a place where there is no eye. All our actual knowledge is concerned with the boundary surfaces between matter and empty space: what is inside and outside these surfaces is conjectural. I cannot help believing that some far simpler logical scheme of physics is possible than any yet evolved, and that the simplification is most likely to come through giving up the attempt to make physical space resemble the space of percepts, of which a beginning has been made by the Heisenberg quantum mechanics. The theory of space-time developed in Chapters XXVIII. and XXIX. was, perhaps, unduly orthodox and unimaginative. Perhaps a great deal of apparatus could be cut away if we could free ourselves from the belief that we must preserve, in physics, characteristics which we find in psychological space and time. To this topic I shall devote the next chapter.

CHAPTER XXXII

PHYSICAL AND PERCEPTUAL SPACE-TIME

In Part II., when we were considering the transition from perception to physics, we took over from common sense certain rough-and-ready approximations which, at our present stage, we must seek to replace by something more exact. We want now to make a second approximation: having inferred a certain kind of physical world from our percepts, we can use the properties of this inferred world to reinterpret the relation of percepts to the outer world, and we can consider more carefully whether any of the properties we assigned to the outer world were accepted without sufficient reason, merely because they were such as we think we find in the perceptual world. The subject is imaginatively difficult, and it is not easy to disentangle different levels of inference, but it is important to do so.

Starting from percepts, we observe that different people have similar percepts, whose differences proceed approximately according to the laws of perspective. The first picture of the physical world to be derived from a comparison of percepts (when we start with a developed logic, not with common sense) is, that there are groups of more or less similar events arranged about centres; that the first-order laws as to the differences between events in one group are spherically symmetrical with respect to the centre of the group; and that the second-order laws are obtained by combining a number of laws of "distortion," each of which has its own centre. In this picture of the world, we use a physical space which is derived from, and also correlated with, the space of percepts, in the manner explained in discussing phenomenalism in Chapter XX. I shall here repeat and amplify this construction, with a view to suggesting modifications of it derived from physics.

We cannot wholly eliminate the subjective factor in our knowledge of the world, since we cannot discover experimentally what the world looks like from a place where there is no one to see it. But we can make the subjective factor approximately constant, and thus be reasonably convinced that the differences which remain are due to causes that are not subjective. I shall therefore suppose that, at a given moment, a number of photographs are taken of some object, say a chair or a table, from different places, with cameras and plates as similar as possible. I shall suppose that the photographs are compared by a person sitting motionless, who places them successively on a fixed stand in front of him. It is then reasonable to assume that the differences between his percepts of the photographs are due to physical causes; also, within limits, that the likenesses between them are due to likenesses in the stimuli to the photographic plates. We find that the differences between the photographs proceed according to certain laws, which we call the laws of perspective; these laws are correlated with the differences between the appearances of the different cameras to an observer who sees them all at the moment when the photographs are taken, and so on. In fact, they can all be expressed as functions of the " co-ordinates " of the cameras and the parts of the table, where " co-ordinates " may be defined by relation to the single observer. *E.g.* he may get another man to go with one end of a stretched tape-measure to each camera in turn, while he holds the other end; he can read the length r of the tape-measure, and observe, on scales, the angular co-ordinates θ, φ of the tape-measure. These facts lead us to attribute a measure of objectivity to our co-ordinates, since, although they are all observed by us from our point of view, they determine the sort of photograph that a camera will take. Further, they lead us to think that, all round the table or chair which is being photographed, there are events which are connected with each other according to the laws of perspective as stated with reference to a certain centre as defined by our polar co-ordinates. Our observer's r, θ, φ are facts concerning his own percepts, yet they suffice mathematically to determine the " percepts " of the cameras; they must therefore have some significance which is not purely private to him.

This argument, elaborated and extended in obvious ways, gives the ground for supposing that our perceptual space has some objective counterpart, *i.e.* that there is some relation between the camera and the table corresponding to the relation between the co-ordinates of our percepts of them. (I am throughout assuming the causal theory of perception.) If we now use one camera to make one photograph containing various objects, we shall again find that the spatial relations of the representations of the objects in the photograph are such as can be calculated from the co-ordinates of the objects and the camera. We cannot know the intrinsic quality of the events at the camera which cause the photograph, but we can infer a certain similarity of structure between these events and our percept of the photograph. All this leads us to the notion of groups of events arranged about centres, the centres having to each other relations whose causal properties can be inferred from relations between certain of our percepts. That is to say, given a group G, of which one member is a percept p, and another group G', of which one member is a percept p', if r, θ, φ are the co-ordinates of p, and r', θ', φ' are the co-ordinates of p', there is a relation between G and G' which can be inferred from r, θ, φ and r', θ', φ'. These facts give the grounds for regarding space as objective, though, even on the basis of these facts, the space which is objective will not be identical with the space of perception, but only correlated with it.

The events which cause a photograph obviously take place

at the surface of the photographic plate; what happens between this and the object photographed consists of causal antecedents, not of the immediate cause. And the resulting photograph is in the plate, not in the object. Similarly the events which are the immediate causal antecedents of our percept are in the eye and optic nerve, and the percept is in us, not in the outer world, when we are speaking of physical space. The whole of our perceptual world is, for physics, in our heads, since otherwise there would be a spatio-temporal jump between stimulus and percept which would be quite unintelligible. Any two events which we experience together—*e.g.* a noise and a colour which we perceive to be simultaneous—are "compresent." I should not say, however, that two percepts which are not both "conscious" *must* be compresent. Two events are compresent when they form together one causal unit or part of one—this is a sufficient, but perhaps not a necessary, condition. When two percepts are experienced together, they are thus causally conjoined; but when either is "unconscious" they may not be, and therefore we cannot be sure that they are compresent. It is not necessary, consequently, to suppose that the mind occupies a mere point in physical space.

It is now necessary to point out the limitations to the accuracy of the above account. In the first place, there are departures from the laws of perspective which can be easily fitted in—opaque bodies, prisms, looking-glasses, echoes, etc. These cases are easy because the departure from regularity as regards one sense is accompanied by evidence, from another sense, of the existence of a physical object at the centre of the disturbance, or at the apex if the disturbance is conical, like a shadow. Then there are the cases where a physical object is inferred from the disturbance, although there is no direct evidence of its existence. But none of these are really important. The two important matters are: (1) The difficulties about measurement; (2) the difference between a percept as it seems and a stimulus as it is inferred.

(1) The difficulties about measurement have already been discussed, but we must now endeavour to reach conclusions about them. As already pointed out, every measurement, however inaccurate, records a fact, though not always the fact which it is intended to record. We saw a moment ago that, if we measure the co-ordinates r, θ, φ of an object to be photographed and of a number of cameras, we can make inferences as to the pictures which the various cameras will make of the object. We inferred that the co-ordinates represented relations to our body which have certain peculiar properties of the sort called geometrical, in the sense that when we know the co-ordinates of two bodies relatively to ourselves, we can infer their co-ordinates relatively to each other. All this is only roughly true if our measurements are careless: in that case, when we mean to discover intrinsic relations, we are only discovering very complicated relations involving our sense-organs and perhaps even our desires. We seek a technique for eliminating all circumstances except those with which we wish to be concerned, and to a great extent we are successful. But relativity informs us that there is a residue of variability in measures which cannot be eliminated, because, in fact, the relations we try to measure are partially non-existent. Or, more correctly, they are relations involving more terms than we thought they did. We supposed that co-ordinates represented relations to the axes. But if we had two sets of axes momentarily coinciding, while one was moving relatively to the other, the co-ordinates of an event would not in general be the same with respect to both. And we cannot even, in any strict sense, discover any exact relation between distant points such as could give physical significance to co-ordinates. The appearance to the contrary is only an approximate truth, which cannot be made precise.

All this represents a failure of correspondence between physical space-time and perpetual space and time. If we assume that the human body moves in a geodesic, perceptual time may be identified with the integral of ds taken along that geodesic, while perceptual space consists of certain relations between simultaneous percepts (the word " simultaneous " raises no difficulties, since all percepts are in our heads), partly themselves perceptual, partly inferred, but all just what they are, whatever physics may say. There are certain respects in which we can modify perceptual space to suit physics, and certain others in which we cannot. We can, for example, infer that percepts consist of imperceptible parts, if physics gives us ground for thinking so. But where we perceive some relation between percepts, we cannot deny that there is such a relation, however little physics may allow it to subsist between the objects said to be perceived. The rule is: We can infer extra complexity of structure in percepts if physics requires it, but, however much physics may require it, we cannot infer a *smaller* complexity than is demanded by the study of percepts on their own account. In the world of percepts, the distinction between space and time does really exist, and space does really have certain properties which relativity denies to physical space. Thus to this extent the correspondence between perceptual and physical space breaks down, and measurement, which has to do primarily with percepts, fails to give us quite such good data as we hoped to obtain for inferences as to the physical world.

(2) I come now to the difference between a percept as it seems and a stimulus as it is inferred. But this is not the whole scope of the problem to be discussed. The word " perception " implies relation to a physical object; we are supposed to " perceive " a chair or a table or a person. If physics is correct, the relation of a percept to a physical object is very remote and curious. In ordinary cases, we see objects by

means of light which is reflected or scattered, which increases the complication. To take the simplest possible case, let us suppose that we are seeing a glowing gas. The percept seems to be a patch of bright colour of a certain shape, sensibly continuous in perceptual space, and approximately constant in perceptual time. Perception gives knowledge only in so far as this percept corresponds to what is really taking place in the gas. Now if physics is true, there are great differences between the apparent structure of the percept and the real structure of what is taking place in the gas. (Differences other than structural may be ignored.) Instead of something steady and continuous, such as the percept seems to be, the process in the gas is supposed to be a large number of separated sudden discrete upheavals. It is true that there are important similarities between the percept and the physical event. The shape of the percept corresponds to the shape of the region in which the upheavals are taking place, with the limitations mentioned just now in connection with measurement. The colour of the percept corresponds to the amount of energy lost by each atom in an upheaval. The constancy of the percept corresponds to the statistical constancy in the rate at which upheavals occur in any not too small portion of the gas. Thus everything in the percept represents a statistical fact about the gas, with the exception of the colour, which is supposed to represent a fact about each atom. This, by the way, is an odd reversal of Locke's dictum about secondary qualities: the colour is the most nearly objective of all the elements in the percept.

These differences are all of one kind in a certain respect: they attribute *more* structure to the physical occurrence than to the percept. This is in line with the general principle that the relation of distant to near appearances is one-many, so that differences in the percept imply differences in the object, but not vice versa. The finer structure of the object is all, in the last analysis, inferred from the grosser structure of percepts, but it involves the comparison of many percepts and the search for invariable causal laws, in the manner which we considered in Part II. There is therefore no inconsistency in the view that the physical event differs from the percept in the way suggested by physics, since the difference consists in attributing more structure to the physical event, not in denying to it those elements of structure which are possessed by the percept.

It is possible, if we choose, to attribute to the percept the same structure as that possessed by the physical occurrence, or rather the same structure as that possessed by the immediate external stimulus. It cannot be proved that this hypothesis is untrue, but it is less useful than it might be supposed to be, because only what is *known* about percepts is epistemologically important, and such structure, if it exists, is certainly unperceived. What we only discover about percepts by means of inference does not belong to the part which affords premisses for science, but is, from the standpoint of theory of knowledge, in the same position as events in the external world. Therefore, although percepts may have an unperceived structure, this does not diminish the significance of the fact that the structure we *perceive* in percepts has only a one-many relation to that of their stimuli.

The question must be faced: Is physical space-time perhaps much more unlike the space and time of perception than we have supposed? Have we been victims of imaginative laziness in our merely piecemeal modifications of common-sense prejudices? Dr Whitehead, most emphatically, is not open to such a charge; his "fallacy of simple location," when avoided, leads to a world-structure quite different from that of common sense and early science. But his structure depends upon a logic which I am unable to accept, namely the logic which supposes that "aspects" may be not quite alike, and yet may be in some sense numerically one. To my mind, such a view, if taken seriously, is incompatible with science, and involves a mystic pantheism. But I shall not pursue this topic here, having treated it on former occasions. The question I wish to ask is: Without adopting heroic measures, what could we suppose about physical space-time, if we were anxious to preserve what is probably true in physics, but not anxious to keep as near as possible to common sense? In particular, can space-time itself be atomic, as the existence of the unit of action h seems to suggest? And first, how are we to conceive "action"?

Action is usually defined as the time-integral of energy; since energy can be identified with mass, "action" may also be defined as mass multiplied by time. Gravitational mass is a length; *e.g.* the mass of the sun is 1·47 kilometres.* Since gravitational and inertial mass are equal, we might regard action as length multiplied by time. Dr Jeans (*Atomicity and Quanta*, p. 8) says:

> "There can hardly be an atomicity of the continuum itself, for, if there were, a universal constant of the physical dimensions of space multiplied by time ought to pervade the whole of physical science. Nothing of the kind is even suspected, nor, so far as I know, has ever been so much as surmised. Thus science can to-day proclaim with high confidence that both space and time are continuous."

In this passage, the "high confidence" seems to me to go beyond what is warranted. If there were a scientific gain in conceiving the space-time structure atomically, I do not believe that any theoretical arguments to the contrary could interpose a veto. Arguments from dimensions, such as Dr Jeans employs, have no longer the definiteness that they had before the introduction of relativity. As we have just seen, we *could* define "action" so that its dimensions would be

* Eddington, *op. cit.*, p. 87.

length multiplied by time. Now there is a universal constant of action, namely *h*. Perhaps, if we were to take action as one of the basic conceptions of physics, we might be able to construct a physics which would be atomistic all through, and yet would contain all that is verifiable. I do not "proclaim with high confidence" that this is possible; I only invite attention to the hypothesis, as worth investigating on the chance of its affording a simplification of the conceptual apparatus of physics. In the following chapters, this hypothesis is to be borne in mind.

CHAPTER XXXIII

PERIODICITY AND QUALITATIVE SERIES

THE periodic character of many physical occurrences has been obvious ever since men observed their own respiration and the alternation of night and day, but it has acquired a quite new importance with the discovery of the quantum. The quantum characterizes a whole period of a rapid periodic process, not any one moment of the period; it thus requires us to consider the period as a whole, and in some sense reverses what has hitherto been the trend of physical laws, namely to proceed away from integrals towards differentials. It will be remembered that the quantum principle, as enunciated by Wilson and Sommerfeld, states: Given a periodic or quasi-periodic process, the kinetic energy of which has been expressed by means of "separated" co-ordinates, if q_k is any one of these co-ordinates and E_{kin} is the kinetic energy, then

$$\int \frac{\partial E_{kin}}{\partial \dot{q}_k} dq_k = n_k h$$

where the integration is to extend over one complete period of q_k, and n_k is a small integer which is the quantum number associated with q_k. This law is essentially concerned with a whole period, and thus makes periodicity fundamental in physics in quite a new way.

Before going further, it will be well to consider how far periodicity retains this importance in the newer quantum mechanics inaugurated by Heisenberg. For this purpose, we may concentrate attention upon the one fundamental equation involving h in the new system. This equation takes the form:*

$$pq - qp = \frac{h}{2\pi i} \mathrm{I}$$

* M. Born and P. Jordan, *Zur Quantenmechanik*, Zeitschrift für Physik, 34, p. 871. Also M. Born, W. Heisenberg, and P. Jordan, *ib.*, 35, p. 562.

where p and q are matrices, q being a Hamiltonian co-ordinate in the new sense, and p the corresponding "impulse," also in the new sense; while 1 is the unit matrix. This equation is asserted to hold for *all* motions, not only for such as are periodic. But in the case of motions which are not periodic, it gives a result which approximates to that of classical mechanics. Thus it remains the case that the new mechanics is only necessitated by periodic motions, although it is technically possible to find a quantum principle which is also applicable to non-periodic motions. Hence the importance of periodicity remains intact from an empirical point of view, though somewhat diminished from the point of view of a statement of fundamental laws. In any case, it remains sufficiently important to demand a separate discussion.

Traditionally, periodicity in physics was a question of motion: a body described the same path in space over and over again. With the coming of relativity, it has become necessary to modify this account somewhat. In space-time, every point has a date, and cannot be occupied twice; neither the earth nor an electron can describe again the orbit it described on a former occasion. And periodicity will be relative to a given system of co-ordinates: if, in one system, a co-ordinate runs through a given range of values repeatedly, and always in equal times, it may happen that, in another system, even if there is an oscillating co-ordinate, its periods are not all equal. A change of axes may even take away all trace of periodic character from a process. Since, however, the quantum principle compels us to treat periodicity as physically important, it would seem that we must regard it as a character belonging to a process when referred to axes which move with it, since this would overcome the difficulties connected with relativity. If, in certain cases, this method is not open to us, some other must be found which equally avoids these difficulties. But where processes connected with matter (as opposed to electromagnetic processes) are concerned we shall, I think, find no other possibility except to take axes which move with the matter concerned. But this makes it impossible to treat periodicity as fundamentally a character exhibited in a motion, since we have reduced to rest the body in which the periodic process is taking place. The suggestion I have to make is that, fundamentally, periodicity is constituted by the recurrence of *qualities*.

In the present chapter, I wish to consider what can be meant by the " quality " of an event; I wish also to investigate the connection of quality with causality and motion and periodicity.

Physics traditionally ignores quality, and reduces the physical world to matter in motion. This view is no longer adequate. Energy turns out to be more important than matter, and light possesses many properties—*e.g.* gravitation—which were formerly regarded as characteristic of matter. The substitution of space-time for space and time has made it natural to regard events, rather than persistent substances, as the raw material of physics. Quantum phenomena have thrown doubt on continuity of motion. For these and other reasons, the old simplicities have disappeared.

When we start from perception instead of from mathematical physics, we find that the events with which we are best acquainted have "qualities," by means of which they can be arranged in classes and series. All colours have something in common which is not possessed by sounds. Two colours may be so similar as to be almost or quite indistinguishable, but they may also be very dissimilar. As *Gestalt-psychologie* has emphasized, shapes are perceived qualitatively, not analytically as a system of interrelated parts. But this whole conception of quality, which plays such a large part in our perceptual life, has been wholly absent from traditional physics. Colours, sounds, temperatures, etc., have all been regarded as caused by various kinds of motions. There was no objection to this so far as it succeeded, but, if and where it proves insufficient, there can also be no objection to re-introducing qualitative differences into the physical world.

There is, however, one essential limitation. We may find reasons for supposing qualitative differences, in order to be able to build up the kind of structure which we have inferred; but we cannot have any means of knowing what are the qualities which differ. This point was discussed in Part II., and need not now detain us.

The apparatus so far assumed, apart from qualities, has been: co-punctuality, cause-and-effect, and the quantum laws. I say "cause-and-effect" because it is necessary to be able to distinguish the earlier from the later event in a transaction, and this is a smaller assumption than that of a general time-order among events in one causal series. The above apparatus sufficed except for one purpose: that of defining "repetition." The possibility of repetition is at the bottom of the common-sense distinction between space and time; the substitution of space-time should, one might suppose, make repetition impossible, and yet the whole of what is distinctive in quantum physics, and the theories of light and sound, not to mention other matters, depend upon periodicity, which involves repetition. So long as we had billiard-balls moving in an unchanging space, we could be content with repetition of configuration. But now spatial distance, which is essential to configuration, has to be analyzed into an elaborate indirect relation depending upon the existence of common causal ancestors or descendants. We must, therefore, be able to distinguish among events by means additional to their space-time relations.

There is, however, a considerable difficulty in finding laws governing what we are calling "qualities." In a world of continuous processes, one would say that qualities must change gradually. But in a quantum process they apparently change suddenly. Perhaps, however, this suddenness does not exist in a steady rhythmic process; or perhaps, even if it does, it may involve small changes producing a serial character in the successive qualities. Take, for example, the revolution of an electron about a nucleus. In the newer quantum theory this does not really occur, but we may consider how it could be interpreted if it were necessary to assume it. Let us make a fantastic hypothesis, purely for illustrative purposes: let us suppose that the electron and the nucleus can see each other, and that neither rotates on its own axis. Then they will get pictures of each other which change during each revolution, and repeat the cycle of changes each time. Now let us turn this hypothesis round, and begin by assuming the recurrent series of pictures. From this we can infer the revolution of the electron, provided we are free to construct space as we like, subject to certain formal laws. Now in fact we have this freedom: the " space " in which the electron revolves need only have certain abstract mathematical properties, and, so long as it has them, it may be constructed out of any material available. So long as the electron continues in one orbit, we may conceive, at any rate as a schematic simplification, that there is a persistent event E which may be taken as representative of it, and in like manner that there is a persistent event P representative of the proton. Now let us suppose that, compresent with E but not with each other, there are successive events p_1, p_2, p_3, . . . which may be regarded as " aspects " of the proton, and are related to each other more or less in the way in which the appearances of the proton from different places would be related if the electron could see. Similarly let us assume a series of events e_1, e_2, e_3, . . . compresent with P but not with each other, analogous to what would be appearances of the electron to the proton if the proton could see. And let us further suppose that, after a certain set of such events, an exactly similar set recurs, or a very approximately similar set. This supposition provides us with the material required for a periodic relative motion. We shall say, therefore, not that perspectives differ because spatial relations change, but that change in spatial relations consists of systematic alteration in perspectives. Such a view is feasible, but it makes similarity and difference of quality essential. It ceases to be fantastic if we drop the analogy with vision except as regards purely formal characteristics.

Let us now set forth the analysis of a periodic process suggested by the above, bringing it into relation with the construction of points in Chapter XXVIII. Let us assume, to begin with, that the process is discrete; this hypothesis can be dropped later, but simplifies the initial statement. Suppose, for the sake of illustration, that there are ten qualities, q_0, q_1, q_2, . . . q_9, and that there exist events

$$a_{10}, a_{11}, a_{12}, \ldots a_{19}, a_{20}, a_{21}, \ldots a_{29}, a_{30}, \ldots$$

which are subject to the following conditions:

(1) a_{10}, a_{20}, a_{30}, . . . have the quality q_0,; a_{11}, a_{21}, a_{31} . . . have the quality q_1, etc.

(2) Each of the a's is compresent with its immediate neighbour to left and right, but with none of the other a's;

(3) If $m < n$, any point of space-time of which a_m but not a_n is a member has a time-like interval from any point of which a_n but not a_m is a member.

In that case, the series of a's constitutes a periodic process, having ten a's in each period. The last digit in the suffix of an a indicates the quality of the a—*i.e.* if the last digit is r, the quality is q_r—while the remaining digits indicate the number of the period.

If all the a's are events in the history of one piece of matter, that piece of matter is undergoing the periodic process. If there is a correlative series of b's in another piece of matter, the two periodic processes together make up one relative motion of a periodic character, such as the revolution of an electron about a proton.

Generalizing the above, while still assuming that the process is discrete, suppose we have r qualities $q_1, q_2, \ldots q_r$, and a set of events

$$a_{11}, a_{12}, \ldots a_{1r}, a_{21}, a_{22}, \ldots a_{2r}, a_{31}, \ldots$$

where, as before, the last suffix indicates the quality, *i.e.* a_{mn} has the quality q_n $(n \leq r)$. Suppose, also, that each a is compresent with p of its predecessors and p of its successors, where $2p+1<r$; but that no a is compresent with any a except these. The remaining assumptions are to be as before. Then again we obtain a rhythm which may be regarded as an analysis of periodic processes in physics.

If we suppose that the a's are not compresent with any events except the other specified a's, then the group of a's with which a given a is compresent constitutes a point, which may be taken as the middle point in the duration of the a in question. We can take this point as representative of the a in question, since their relation is one-one. Thus the a in question is associated with a point, in spite of the fact that it lasts for a finite time, *i.e.* is compresent with events not compresent with each other.

It is to be observed that, according to the theory of space-time in Chapters XXVIII. and XXIX., it is quite possible for some parts of space-time to be continuous and others discrete. I am supposing, at the moment, that we are considering a periodic process in a discrete part of space-time; this does not involve the hypothesis that *all* space-time is discrete.

If the a's in one periodic process, as we supposed a moment ago, are not compresent with any events except certain neighbouring a's (which must be fewer than the whole of one period), then the number of points in a period is the same as the number of a's, and either affords a measure of the duration of the period, measured by its proper time. It is obvious that, in a discrete part of space-time, the natural measure of distance will be number of intermediate points. We see also how the proper time of one process can differ from that of another. Let us suppose that our a's form an "isolated" process (*i.e.* are not compresent with anything except other a's), except at the beginning and end; the first and last a's are to be compresent with the first and last terms of another periodic process composed of b's, which also is to be isolated except at its ends. Then the proper time of the b-process is measured by the number of b's between the two ends, which need not have any relation to the number of a's. This illustrates, what of course follows from relativity, that periodicity must be measured by standards intrinsic to the process concerned, not by standards appropriate to other periodic processes. Such remarks would hardly be necessary but for the fact that relativity and quantum theory at present stand apart from each other, and have not yet been brought into one whole by the physicists.

The above can be stated in the language of mathematical logic, thereby making the character of the assumptions clearer and the generalization to continuous processes easier. Let Q be the series of qualities, A the series of events in the rhythmic process. Let us imagine the events arranged in rows and columns, so that each row consists of one period and each column consists of all the events having a given quality. We assume a one-many relation S, whose domain is the field of Q and whose converse domain is the field of A. When q has the relation S to a, we say "a has the quality q." If a is any term in the field of A, let q be the term which has the relation S to a; then the next term below a in the same column (*i.e.* the corresponding a in the next period) is the first term a' in the A series which is after a and to which q has the relation S.

The " row of a " consists of all a's earlier than a' and not earlier than a. The " column of a " consists of all a's to which q has the relation S. We assume that S with its converse domain limited to one row is one-one, so that each row (*i.e.* each period) is a series which is similar (in the technical sense) to the series Q.

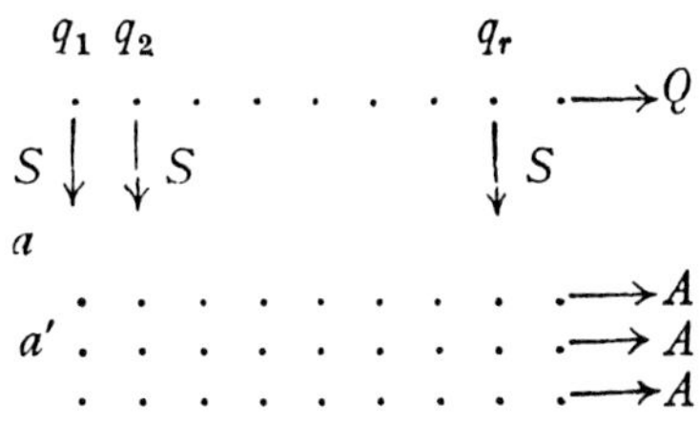

There is no difficulty in adapting the above analysis of periodicity to continuous processes. Instead of an enumerated set of qualities $q_1, q_2, \ldots$, we shall have to take some continuous series of qualities, such as the colours of the rainbow, or the notes produced on a violin by running one's finger up and down the string. The number of events compresent with a given event must now be infinite, but must still be less than the whole of one period (ignoring events outside the process concerned). The number of points in one period, or in any finite portion of it, is now infinite, and cannot therefore be used as a measure of distance. Thus in regard to metrical properties there are important differences between continuous and discrete processes. However, I shall not enlarge upon these, as I propose to consider the analysis of " interval " in a later chapter.

Hitherto I have been considering processes which may be regarded as taking place in matter, or which, at any rate, do not move with the velocity of light. But light, also, is commonly regarded as consisting of a periodic process. Accepting the wave-theory of light, let us proceed to analyze its periodic character. We shall find that it differs in important respects from that of periodic processes in matter.

The periodic character of a light-wave cannot exist from its own point of view, but only from that of the matter which it encounters or from which it radiates. We may suppose that when light radiates from an atom at the time of a quantum change, there is, from the point of view of the atom, a temporal series of what we may call "luminous events," and that this series is periodic in the sense which we have been considering. One period of such luminous events constitutes the emission of one light-wave. If we suppose that the light is absorbed by another atom, we may suppose that each of the luminous events is compresent with certain events in the absorbing atom, as well as with certain events in the emitting atom. As measured by the proper times of the atoms, the time-order of the luminous events is the same for the two atoms. But from the point of view of the luminous events themselves, there is no periodicity. So long as the light does not encounter matter, it consists of separated events which at most "touch" one other event at each boundary; the traveller who accompanies one of the events can have no cognizance of any of the other events, since they cannot catch each other up. If we could imagine a homunculus floating on the crest of a light-wave, he would have no means of discovering that anything periodic was occurring, since he could not "see" the other parts of the wave. The different parts of a light-wave cannot, in a word, interact causally in any way, because no causal action can travel faster than light.

We cannot even properly speak of a periodicity in the light-wave for an observer who watches it pass. We can only see light by stopping it. This applies to such phenomena as interference, which is only made visible by allowing light to meet matter. It is true that interference gives us a ground for inferring structure: two processes can neutralize each other, but two "things" cannot. If *A* owes *B* a pound, and *B* owes *A* a like sum, the result is zero; but if *A* has a pound in his hand to give to B, and B has a pound in his hand to give to A, there are two pounds. Wherever the sum of two occurrences can be null, both occurrences must have a relational character. Thus we are justified, by such facts as interference patterns, in supposing that, when light falls on a body, the body experiences a series of events whose effects upon it are of opposite kinds, as if some pushed it one way and some another. But all this is from the point of view of the body, not of the light. Thus the frequency of light is a characteristic which exists for a body which emits light, and for a body which absorbs it (*e.g.* the body of a scientific observer), but not for the light itself while it is *in vacuo*.

When light is emitted and absorbed, we may therefore suppose that what happens is according to the following scheme. We have a temporal series of events in the emitting body, and, compresent with each of these, a luminous event. These luminous events, arranged in the time-order of the compresent events in the emitting body, form a periodic process in the previous sense. Each of the luminous events is also compresent with some event in the absorbing body. The time-order of the events in the absorbing body is the same as that of the events in the emitting body; *i.e.* if e, e' are events in the emitting body, compresent respectively with l, l', two luminous events; and if l, l' are respectively compresent with a, a', two events in the absorbing body, then if e is earlier than e', a is earlier than a'. What happens to light-waves which are emitted but not re-absorbed we cannot tell, since, by the nature of the case, there can never conceivably be any evidence on the point.

According to the above, the frequency of a light-wave is a characteristic which it has in relation to matter, not in relation to itself. In this it differs from, *e.g.*, the periodicity in the revolution of an electron, which may be supposed to exist for the electron itself.

The chief point of the above hypothesis is the suggestion that single "luminous events" extend from the emitting to the absorbing body. I do not advance it as anything more than a possible hypothesis. One of its main purposes is to account for the fact that the interval between two parts of a light-ray is zero; but this part of the argument belongs to a later chapter.

CHAPTER XXXIV

TYPES OF PHYSICAL OCCURRENCES

In this chapter, I propose to advocate a division of physical occurrences into three types, which I shall call respectively steady events, rhythms, and transactions. The phrase "steady events" is formed on the analogy of "steady motions," though the events concerned are not supposed to be motions. Rhythms are periodic processes, such as we considered in the preceding chapter. Transactions are quantum changes, in which energy passes from one system to another. The laws governing different types of occurrences are different, and it is necessary to separate them before embarking upon a general discussion of physical causality.

The traditional view, that physics is concerned exclusively with matter in motion, cannot be maintained, for a number of reasons. In the first place, the æther, even if it can be said to exist, can hardly be regarded as having a granular structure, and events in it, such as the passage of light, cannot be explained as movements of particles of æther. In the second place, quantum changes, if they really are sudden, violate the continuity of motion, and thus destroy its advantages as an imaginative picture. In the third place—and this is philosophically the most important point—the conception of motion depends upon that of persistent material substances, which we have seen reason to regard as merely an approximate empirical generalization. Before we can say that one piece of matter has moved, we must decide that two events at different times belong to one "biography," and a "biography" is defined by certain causal laws, not by persistence of substance. Consequently motion is something constructed in accordance with the laws of physics, or—we might say—as a convenience in stating them; it cannot be one of the fundamental concepts of physics. Lastly, there is an argument which is difficult to state precisely, but which nevertheless has some weight. For Newton, motion was absolute, and a body in motion might be regarded as in a different state from a body at rest. But when motion was recognized as merely relative, laws of motion became laws as to relations to more or less distant bodies. They thus came to involve something like action at a distance—though this was disguised by the use of differential equations not always interpreted according to rigid Weierstrassian methods. If we are to avoid action at a distance, our fundamental laws must be concerned with terms having finite spatio-temporal extension, and thus capable of contact and overlapping—in a word, with events rather than particles or impenetrable material units. This involves a re-interpretation of motion as it occurs in physics, which will be considered in a later chapter. For the present, I am concerned with the materials which will be required for this purpose as well as for the interpretation of other physical phenomena.

A "steady event," as I use the term, is anything which is devoid of physical structure and is compresent with events which are not compresent with each other, but are one earlier and the other later; in other words, the steady event is a member of at least two points which have a time-like interval. When a steady event is contrasted with a rhythm, it is assumed that the steady event is not part of a periodic process; but it cannot be taken as certain that there are any elementary events which are not parts of such processes. It may be that all non-periodic changes occur by way of transactions; but this must be an open question in the present state of knowledge.

A "rhythm," as already explained, is a recurring cycle of events, in which there is a qualitative similarity between corresponding members of different periods. A rhythm may have a period consisting of a finite number of events, or one consisting of an infinite number; it may be discrete or continuous. If it is discrete, the proper time of one period is measured by the number of events in the period, and the "frequency" of the process is the reciprocal of this number. But here we are speaking of the frequency as measured by the time proper to the period; by an extraneous time the frequency may be quite different. What is commonly called the frequency of a light-wave is its frequency with respect to axes fixed relatively to the emitting body. Its frequency relative to axes which travel with it is zero; this is only the extreme of the Doppler effect. There is perhaps a certain inconsistency in the practice of studying bodies by means of axes which move with them, while light is always treated with reference to material axes. If we want to understand light in itself, not in its relation to matter, we ought to let our axes travel with it. In that case, its periodicity is spatial, not temporal; it is like that of corrugated iron. From the standpoint of the light itself, each part of a light-wave is a steady event in the sense defined above.

One of the most fundamental of rhythmic processes will be the revolution of an electron about a nucleus, unless we accept the view of the new quantum mechanics, according to which there is no reason to suppose that this really occurs. In the Bohr-Sommerfeld theory, this revolution goes on by itself until it is altered either by a quantum change or by some more conventional chemical or electrical action. The question arises: why should we suppose that there is a process at all? Why not suppose that there is a steady event, possessed of a certain amount of energy, which is replaced, in a quantum change, by another steady event, possessed of a different amount of energy, the balance being radiated or absorbed? There is a certain attraction about this hypothesis, since the atom gives no external indication of its presence while the supposed process continues, and therefore there can be no direct evidence that changes are occurring, such as a steady motion supposes. In any case, if an electron is revolving round a proton in a circle, and both are spherically symmetrical, it is not easy to see, from a relativist point of view, exactly what is meant by saying that the electron is revolving. This difficulty is not diminished by the hypothesis of spinning electrons. We have the same difficulties as in the case of absolute rotation and Foucault's pendulum—the difficulties, namely, which Newton advanced to prove the necessity of absolute motion. Within the system consisting of the electron and proton alone, nothing is changing while the electron revolves in its circular orbit; the change is only with reference to other bodies. Why not regard the state of affairs as static, but possessed of a certain amount of energy? Energy may be altered in amount by a change of axes, and is not an invariant property of the system; but reference to the outside world here is less serious, since the only purpose served by the energy of the atom is to provide physics with something which can be radiated into the outer world or absorbed from it. That is to say, energy is required only as something whose changes govern the causal relations of the atom with the outer world. This point of view is essentially that of the Heisenberg theory.

There are several apparent difficulties in such a view. In the first place, the formula for energy obtained on the assumption that the electron revolves gives exactly the changes of energy required to account for spectroscopic phenomena; the Bohr-Sommerfeld theory agrees with observation so minutely that its formula for energy must be accepted. Of course we could say that the energy just happens to be what it would be if the electron revolved in one of the quantum orbits; but this would seem an almost miraculous coincidence. This, however, is not the strongest argument, which is that derived from the quantum principle. The quantum principle in its older form can only be applied to periodic processes; if it is to apply, as we find that it does, to the interchange of energy between light and the atom, we must assume, if we adhere to the older theory, that within the atom there is something that can be called a "frequency," *i.e.* something which is periodic, which compels us to admit that, within an atom in a steady state, there is a recurring process whose formal properties are those which would be exhibited by a revolution of the electron, and perhaps also by a rotation.

If we adhere to the Bohr theory, what can be supposed to be really occurring? If relative motion were all that was taking place, we should have either to find an interpretation for the spinning electron, or else to say that, taking axes fixed relatively to any large body, the line joining the electron to the proton rotates rapidly; any large body will do, since none rotates with an angular velocity comparable to that of the electron. But why should the electron be interested in this fact? Why should its capacity for emitting light be connected with it? There must be something happening where the electron is, if the process is to be intelligible. This brings us back to Maxwell's equations, as governing what is occurring in the medium. And there must be a rhythmic character in the events occurring where the electron is, if we are to avoid all the troubles of action at a distance.

We suppose, therefore, that throughout an electromagnetic field there are events whose formal properties we know more or less, and that they, not the change of spatial configuration, are the *immédiate* causes of what takes place. This brings us back to the cycle of events which we used in the preceding chapter to define a rhythm. The point is that a rhythm can never consist merely in periodic changes of spatial relation between two or more bodies, but must consist of qualitative cycles of events. We have experience of such cycles when we watch a large-scale periodic event, such as the swing of a pendulum. All that happens *to* us during the cycle happens *in* us, not in a number of different places; and any effect upon us depends upon what happens to us. I am suggesting that this is a proper analogy when we wish to understand how a periodic motion affects an electron.

I come now to what I shall call "transactions," by which I mean quantum changes. I call them "transactions" because energy is exchanged between different processes. The processes concerned must be periodic, since otherwise the quantum principle is unnecessary. In the simplest case, that of emission of light by a hydrogen atom, we have as antecedent, speaking the language of the older quantum theory, one periodic process (the revolution of the electron in an orbit other than the minimum orbit) and as consequent two processes, namely: (1) The revolution of the electron in a smaller orbit, (2) a light-wave. The latter, as already explained, is only periodic in a certain sense. The energy of the antecedent is the sum of the energies of the consequents. The amount of action during one period of the antecedent is a multiple of h, and so are the amounts of action of the consequents during one period of each. Exactly the converse occurs when light is absorbed by a hydrogen atom. In other cases, both the antecedent and the consequent may consist of two or more rhythms; but always there will be conservation of energy, and each rhythm will contain an amount of action which is a multiple of h.

As yet, everything concerned with quanta is more or less mysterious, although Heisenberg's theory has somewhat diminished the mystery. We do not know whether quantum changes are really sudden or not; we do not know whether the space concerned in atomic structure is continuous or discrete. If electrons always moved in circles, as in the first form of Bohr's theory, we could be content with a granular discrete space, and suppose that the intermediate orbits are geometrically non-existent. But the existence of elliptic orbits in Sommerfeld's development of the theory makes this difficult. And in atoms with many planetary electrons, the paths of some are supposed to cross those of others. In spite of these difficulties, however, I do not despair of the hypothesis that space-time is discrete. The older quantum theory uses the traditional conceptions of physics, and thinks of geometrical orbits in a constant space. The Heisenberg theory, on the contrary, has a completely new kinematics, according to which unquantized orbits (if we may still speak of orbits) are geometrically impossible. It is difficult, as yet, to translate this theory out of its technical form. But even according to the older theory, one can see that a discrete space-time is possible. For when we think of the matter in terms of space-time, we realize that the geometry of the neighbourhood of the atom may be different at different times. If an electron moves in one sort of orbit at one time and in another at another, it does not follow that each sort of orbit was geometrically possible at the time when the other was being described. Perhaps it is not superfluous to explain what is meant by saying that an orbit is "geometrically possible" though not physically actual. What is meant is this: there is a series of groups of events, each group being a point, and the series being one in which all the intervals of points are time-like, and in which, if a constant value is assigned to one of the co-ordinates, the remaining three give a curve in a three-dimensional space having the geometrical properties of the orbit in question. Whenever we speak of an orbit geometrically, we are assuming that we can distinguish one of the co-ordinates as "time," give it a constant value, and consider the relations of the remaining three co-ordinates. Now it is always possible that there may be a fallacy in this procedure, since it may be that such geometrical relations as we are considering are impossible among "simultaneous" points. Moreover, in the general theory of relativity, it may be impossible to distinguish one co-ordinate as more representative of time than the others.

When, from a traditional point of view, two orbits cross each other, this no longer happens from a relativity standpoint. We cannot assume, that is to say, that there is a point from which two journeys are possible. Two electrons never actually collide. When their orbits are said to cross, all that is meant is this: In the system of co-ordinates we have adopted, there is a point (x, y, z, t) which is part of the history of one electron, and a point (x, y, z, t') which is part of the history of the other. In another equally legitimate system of co-ordinates, these two points would not have three co-ordinates identical. And the fact that a certain orbit passes from (x, y, z, t) in a certain direction does not imply that there is an orbit passing from (x, y, z, t') in a direction which is the same so far as x, y, z are concerned. Therefore the apparent difficulties in the way of a discrete space are not necessarily insuperable.

From our point of view, it is a difficulty in the quantum principle that it is stated in a form involving energy, which, from a relativity standpoint, requires reinterpretation. It is also a difficulty that we do not know any laws determining *when* a transaction will take place, and that we do not know whether it is really sudden or not. For all these reasons, we are compelled to be very tentative in philosophizing. I will, however, repeat the outcome of this chapter, such as it is.

In one sense, the theory of space-time points as groups of events requires that all change should be discontinuous. An event e is a member of a certain set of space-time points, and of no others: the boundaries of the region constituted by this set are the boundaries of e, so that it comes into existence suddenly and ceases to exist suddenly. Nevertheless, we can, if necessary, provide for continuity within this scheme.

Suppose a continuous series of qualities, like the colours of the rainbow; suppose that, in some process, each of these is compresent with its neighbours up to a certain distance in either direction, but not with more distant members of the series. Then the group of qualities existing at a point will change continuously, although each separate quality changes discontinuously. We may suppose this to be the nature of change between transactions, and in particular during a rhythm. There is no proof that change is ever continuous, but there is also no proof that it is not. We will assume, for the moment, that change between transactions is continuous in the above sense, but that transactions are discontinuous. This assumption is made only for the sake of brevity of statement; it is not asserted to be true, or even more probable than the opposite assumption.

If we take the above view, there will be three kinds of things to consider in physics: transactions, steady events, and rhythms. Transactions are dominated by quantum laws. Steady events continue, without internal change, from one transaction to the next, or throughout a certain portion of a continuous change; percepts are steady events, or rather systems of steady events. The relation of a steady event to a rhythm I conceive according to a musical analogy: that of a long note on the violin while a series of chords occurs repeatedly on the piano. All our life is lived to the accompaniment of a rhythm of breathing and heart-beating, which provides us with a physiological clock by which we can roughly estimate times. I imagine, perhaps fancifully, something faintly analogous as an accompaniment to every steady event. There are laws connecting the steady event with the rhythm; these are the laws of harmony. There are laws regulating transactions; these are the laws of counterpoint.

We must assume periodicity as a feature of the state of affairs where there are steady events, since we cannot state the quantum principle without it. We have to find a meaning for "frequency" in order to connect energy with h. It is not altogether easy to see how one frequency is to be compared with another. In the case of light, we can estimate the distance between the crest of one wave and the crest of the next. Knowing the velocity of light, this tells us how many waves pass a given place in a second. But here the periodicity exists for the outside observer; for an observer travelling on the crest of a given wave, there is no process and no periodicity. For an outside observer, there is a process in the motion of the light-wave; but our observer on the wave considers himself to be at rest, and presumably does not see objects flying past him. Thus for him the periodicity of a light-wave is spatial rather than temporal. One light-wave will consist of a series e_{11}, e_{12}, e_{13}, . . . e_{1n}, . . . of steady events, the intervals between which are space-like; the next will consist of a series e_{21}, e_{22}, . . . e_{2n}, . . ., again having space-like intervals from each other and from the previous series; e_{1n} and e_{2n} will have a similarity of quality which neither has to e_{1m} or e_{2m} (where m is different from n). Each of these events is supposed to continue as long as the light-wave continues, *i.e.* until there is a transaction. Given any event e which is connected with matter, e may be compresent with e_{11}, e_{12}, . . . e_{1n}, . . . e_{21}, e_{22} . . . e_{2n}, . . . successively, but not with all at once. This is what happens when a light-wave passes an observer or any other piece of matter. A series of events forming one light-wave are inseparably associated, in the sense that when there is one of them there will be others throughout the space covered by the wave. Similarly the series of events (if any) involved in the revolution of an electron are inseparably associated; but there is this difference, that these events form a temporal series from the standpoint of the electron, whereas the events constituting a light-wave form a spatial series from the point of view of the light-wave.

There are difficulties in the above which might be resolved in various ways, but we do not know which to choose. What, for example, shall we say about the transaction which consists in the absorption of energy by an atom from a light-wave? The correct view is supposed to be that, in such a case, a planetary electron passes suddenly from a smaller to a larger orbit. But if we imagine a light-wave to consist of a number of events $e_{11}, e_{12}, \ldots e_{1n}, \ldots$, one might expect that at least one whole wave would be required to produce one definite effect, and that a part of the wave would produce only part of the effect, if any. But a whole wave takes a finite time to reach the atom. This difficulty exists for any view which regards light as consisting of waves and quantum transitions as sudden, but would be obviated if either of these suppositions were dropped. We may therefore take it as part of the general unsolved problem of the relation between radiant energy and energy associated with matter. This problem, though it interests the philosopher, belongs to the domain of physics, and can only be profitably considered by a physicist. I am therefore content to await the discoveries of others.

As regards quanta, let us examine once more what is implied by the fact that there is an important constant h. In the first place, h only exists, or at any rate is only important, in the case of periodic processes, and it is a characteristic of one complete period. In the second place, only integral multiples of h occur. In the third place, when a transaction involves the loss by one system of a certain multiple of h, another system may acquire another multiple of h: what is transferred always unaltered in amount is energy. These seem to be the most significant facts about h.

It seems impossible to resist the view that h represents something of fundamental importance in the physical world, which, in turn, involves the conclusion that periodicity is an element in physical laws, and that one period of a periodic process must be treated as, in some sense, a unit. This follows from the fact that processes arrange themselves so as to secure that a period shall have an important property. This property is simplest in the case of a light-wave: the energy of one light-wave multiplied by the time it takes to pass a given material point is h. If we take the velocity of light as unity, the time a light-wave takes to pass a given point is equal to the spatial distance between the beginning and end of the wave; therefore this distance multiplied by the energy is h. This form might seem preferable for our purposes, since it does not involve reference to an extraneous material point. At least, it does not *obviously* involve such reference; but perhaps the reference is concealed in the process of estimating spatial distance. We have seen that this process must be indirect; one part of a light-wave cannot catch up another, so that the space-like interval between them can only be estimated by means of some process taking place in matter.

If it should be found that quantum phenomena are not physically fundamental, much of what has been said in this chapter will become unnecessary. It should be said, however, that relativity should prepare our minds for the oddest feature of the quantum theory, namely the existence of causal laws involving whole periods. The causal unit, on relativity principles, should be expected to occupy a small region of space-time, not only of space; it should not therefore be instantaneous, as in pre-relativity dynamics. If we combine this with the hypothesis of a discrete space-time, we can imagine a theoretical physics which would make the existence of the quantum no longer seem surprising.

I have to confess, reluctantly, that the theory developed in the present chapter, inadequate as it is, is the best that I know how to suggest on the topic of quanta. Perhaps the progress of physics will make a better philosophy of the subject possible before long. Meanwhile I commend the matter to the consideration of the reader.

CHAPTER XXXV

CAUSALITY AND INTERVAL

THE conception of "interval," upon which the mathematical theory of relativity depends, is very hard to translate, even approximately, into non-technical terms. Yet it is difficult to resist the conviction that it has some connection with causality. Perhaps a discontinuous theory of interval might diminish the obstacles to such an interpretation. Let us try to discover whether this is the case.

The view which naturally suggests itself as a point of departure is something like this: Given two groups of co-puncual events, it may happen that at least one member of one group has a causal relation to at least one member of the other group; in that case, the interval between the two groups is time-like. If causality is a matter of discontinuous transitions, one might expect that the magnitude of the interval would be measured by the number of intermediate transitions. Again, it may happen that no member of one group has a causal relation to any member of the other, but that both contain members having causal relations to a member of a third group. In that case, the interval will be space-like, and again one might suppose that the number of intermediate links would determine the magnitude of the interval.

This represents what might be hoped, but as it stands it is unduly simple, and open to obvious objections. Let us see, therefore, whether it is possible to answer the objections, or to introduce such modifications as will obviate them.

First, let us be clear as to what we mean by a causal relation. There is a causal relation whenever two events, or two groups of events of which one at least is co-punctual, are related by a law which allows something to be inferred about the one from the other. Formerly, one would have supposed that everything about the later event could be inferred from a sufficient number of antecedents; but in view of the explosive and apparently spontaneous character of radio-activity and quantum changes, we must be content with a more modest definition so far as this point is concerned. In another respect, however, our definition is less modest than it would formerly have been. In classical dynamics, causal laws connect accelerations with configurations, so that from the present state of a small region we cannot accurately infer anything as to what will be happening there after a finite time. Quanta have altered this: we can associate the light radiated from an atom with its causal origin, until it hits other matter; we can associate the state of the atom after the emission of the light with its state before, until it undergoes another quantum change. In fact, as we saw in the preceding chapter, we can analyze the course of nature into a set of steady events and rhythms with causal relations governing the "transactions" in which rhythms undergo changes. The above definition was framed with these considerations in mind.

We shall say, then, that all causal relations consist of a series of rhythms or steady events separated by "transactions." If such a series connects a rhythm or steady event A with a rhythm or steady event B, we shall say that A is a "causal ancestor" of B, and B is a "causal descendant" of A. We may assume that, in such a case, the number of transactions between A and B is always finite, since one supposes that the time between two transactions cannot fall below a certain minimum, or at any rate that the number of causally connected transactions in a finite time is never infinite. Perhaps we may assume that a rhythm must last long enough to achieve an amount of action h; perhaps, even, we could construct a discrete theory of time from which this result would follow. All this, however, is very speculative.

Now let us consider the stock case of a light-signal sent from *A* to *B*, and reflected back from *B* to *A*. Only two transactions are involved, namely the emission and reflection of the light; perhaps we ought to add the final transaction, namely the re-absorption of the light by *A*. In any case, there need be only two steady events, one in the outward beam and one in the returning beam. But the interval between the departure and return of the light may have any magnitude. This is all the more curious, as the interval between the departure of the light from *A* and its arrival at *B* is zero, and so is the interval between its departure from *B* and its return to *A*. This suggests that too much effort has been made to regard interval as analogous to distance in conventional geometry and time in conventional kinematics. Suppose we say that, if an event e_1 is a causal ancestor of an event e_2, we take all the possible causal routes from e_1 to e_2, and choose that which contains the greatest number of events: then the "interval" from e_1 to e_2 is defined as the number of events in this longest route. It is obvious that, if a measurable time elapses between the departure of the light from *A* and its return to *A*, there must have been a variety of events at *A* meanwhile. When I say "at" *A*, I have a meaning to be considered shortly; but for the moment it is enough to say that this meaning includes causal inheritance. Thus we have a meaning for the view that the interval at *A* is quite long, and also for the view that the interval between the departure of the light from *A* and its arrival at *B* is zero. This latter statement means that it is the very same event that starts from *A* and arrives at *B*, and moreover that there is no longer causal route connecting the two transactions of starting from *A* and arriving at *B*. This event which starts from *A* and arrives at *B* I call a "luminous event."

But we must deal with space-like intervals before we can decide whether the above theory of time-like intervals will do. It is to be observed that space-like intervals are obtained by calculation from time-like intervals. Let us imagine the following ideal experiment: An astronomer on the sun sends a message to an earthly mirror, and an astronomer on the earth sends one to a solar mirror. Each observes the time of departure and return of his own message, and the time of arrival of the other's message. Each finds that the other's message is received at a time half-way between the arrival and departure of his own message. They compare notes, and discover this fact about each other's observations. They will conclude that, according to the reckonings of both, the two messages were despatched simultaneously, and that the measure of the space-like interval between the despatch of the two messages is half the time between the despatch and return of either, *i.e.* about eight minutes. We may re-state the general method involved as follows: Let us have two transactions S and T connected by a number of causal routes, all going straight from S to T; and let the longest of these consist of n events. Suppose that there is another transaction S' such that its later event extends to T, and that there is no longer causal route from S' to T, nor any causal route at all from S to S'. Here S corresponds to the sending of the signal from the earth, S' to the sending of the signal from the sun, and T to the arrival of the solar signal at the terrestrial observatory. The question is: What is to be the interval between S' and S? There cannot be a causal route from S' to S, because if there were it could be prolonged to T, and would be longer than the single event which extends from S' to T, *contra hyp*. Thus no causal series connects S and S'; there is a causal series connecting S and T; and S' is a transaction that begins an event which ends in the transaction T. In these circumstances, we say that the interval between S and S'

is of a different kind from that between S and T, but has the same numerical measure. The fact that this definition works is what appears as the constant velocity of light.

Difficulties, however, still suggest themselves. What are we to do with the bending of light in a gravitational field? And what are we to say about the connected theory, according to which the velocity of light *in vacuo* is not strictly constant? We have been attempting to regard the passage of light from one body to another as a single static occurrence, involving no change within itself, and therefore having zero for its proper time, since time must be measured by changes. If we have to suppose that the light from a star alters its direction as it passes near the sun, we shall have to think of the journey of the light as a process, not as a mere continuing event. I do not believe, however, that this would be regarded as the correct account of the influence of gravitation on light. Gravitation consists in the fact that a geodesic is geometrically different from what it would be in the absence of a gravitational field; the course of the light is not "really" bent, but is "really" the straightest course geometrically possible. In any case, this point arises at an advanced stage in the theory of relativity, and the considerations involved are so numerous that it would almost certainly be possible to find an interpretation consistent with our suggestion if no other obstacle existed.

When an interval is space-like, it is always theoretically possible to send a light-signal from one of the events concerned to a causal descendant of the other; consequently our definition of the measure of a space-like interval is always possible.

To say that the greatest velocity in nature is that of light is to say that, when two transitions are the beginning and end, respectively, of one luminous event, there is no transition which is a causal descendant of the one and a causal ancestor of the other. To say that a causal chain of transitions belongs to the history of one piece of matter is to say that no two members of the chain can be connected by a chain longer than the portion of the given chain which lies between the two transitions. This is our translation of the law that the history of a piece of matter is a geodesic.

The fact that the interval between two points of one light-ray is zero appears, on the above theory, to be just what might be expected. For when an event has temporal extension, that means that two events which are compresent with it have a causal relation to each other; while when an event has spatial extension, that means that two events compresent with it have a common causal ancestry or posterity. Neither happens in the case of a luminous event, which therefore has neither temporal nor spatial extension, in spite of the fact that it covers a whole region of space-time points.

It will be seen that, according to the above, intervals are discrete, and are always measured by integers. There is, so far as I know, no empirical evidence for or against this view. If the integers concerned were very large, the phenomena would be sensibly the same as if intervals could vary continuously. I do not put forward the theory with any confidence in it as it stands, but rather to suggest to men with more physical competence the possibility of great changes in our picture of the world without rejecting anything probably true. In order to bring out this point, I shall now re-state the theory without interposing argumentative justifications.

The world, it is suggested, consists of a number of events, each involving no change within itself, but each connected with earlier and later events by quantum or other laws which enable us to regard the earlier as the cause and the later as the effect. The quantum transition I call a "transaction." A transaction is subject to laws as to the conservation of energy and as to action. Events may be compresent, and one event may be compresent with a number of others which are separated by transitions; in that case, the one event is said to last for a long time. We can even obtain a continuous time in our theory, if the number of events compresent with a given event is infinite, and their beginnings and ends do not synchronize, *i.e.* one of them may be compresent with two others which are not compresent with each other. But I see no reason to suppose that the number of events compresent with a given event is infinite, or to desire a theory which makes time continuous; I therefore lay no stress upon this possibility.

In a transaction, or during a rhythm, the causal antecedent may consist of more than one event, and so may the causal consequent; but the events which constitute the causal antecedent must all be co-punctual, and so must those which constitute the causal consequent. Any event of the antecedent group will be called a "parent" of any event of the consequent group. When two events are connected by a chain of events, each of which is a parent of the next, the one is said to be an "ancestor" of the other, and the other a "descendant" of the one. Two events may be connected by many causal chains, but all will consist of a finite number of events, and we assume that, in the case of any two given events, there is a maximum to the number of generations in the various lines of descent connecting them. This maximum number is the measure of "interval" when the interval is time-like. When the interval is space-like, the definition of interval is slightly more complicated.

To define space-like intervals, we must first say a few words about light. When a luminous event travels from one body to another, I regard the whole as one static event, involving no internal change or process. Consequently, from the standpoint of the event itself, if one could imagine a being of whose biography it formed a part, there is no time between the beginning and the end. Since nothing travels faster than light, it is impossible that two parts of one luminous event should be compresent with two events of which one is a causal descendant of another; therefore there is no extraneous source from which the luminous event can discover that it is lasting a long time, and there is, in fact, no meaning in saying that it is lasting a long time. But when we say that it is reflected back to its starting-point, we mean that it has undergone a transaction which has turned it into a new luminous event, and that this new event is compresent with causal descendants of events compresent with the earlier one, these compresent events being not luminous, but of the kinds associated with matter. Now, given any two events S and S', neither of which is an ancestor of the other, it is possible to find a luminous event compresent with S' and with a descendant T of S. We then say that the events S and S' have a space-like separation, whose measure is that of the time-like separation between S and T.

In the above theory, it is assumed that, in all cases where one process or piece of matter has an effect upon another, there is at least one event which is compresent with both. This is the form taken by the denial of action at a distance.

If we assume, as we have been doing, that change is discontinuous, a single period of a rhythm will contain some finite number of points. Suppose, now, that there are two rhythms such that the initial event of a period in the one is always identical with the initial event of a period in the other, but the other events are diverse; and suppose that the first rhythm contains m event in a period while the second contains n. Then a period of the first rhythm will contain m points, and one of the second will contain n. We said that the " interval " between two events was to be the number of points in the longest causal route from one to the other; hence the interval between the beginning and end of a period in *either* rhythm is measured by the greater of the two numbers m and n. Suppose this is n. Then we may regard the m-rhythm as having a smaller " velocity " than the n-rhythm, while the frequencies of the two rhythms would be the same. This suggests, in a certain class of cases, a possibility of defining "velocity" otherwise than by relative motion. How far the resulting properties of "velocity" would resemble those resulting from the usual definition, I do not know.

There is no difficulty in defining what is to be meant by saying that a steady event "moves." An event E occupies a number of points of space-time, which can be regarded as a four-dimensional tube divisible into sections such that all the points in one section are simultaneous, and are all later or all earlier than all the points in another section. We shall then regard our event E as moving along the tube, and occupying the various instantaneous sections successively. But this does not imply any process or change within E; it merely implies transitions among events compresent with E but not all compresent with each other. It seems, therefore, that everything essential to theoretical physics can be stated in terms of our theory.

According to the above theory, motion is discontinuous. But this hypothesis is required for one purpose only, namely for the definition of interval. It is easy to introduce such axioms as shall make our space-time continuous, and secure, as in current physics, that discontinuity shall be confined to quantum phenomena, *i.e.* to what we have called "transactions." But if this is done, our definition of interval must be abandoned, and interval resumes its place as something mysterious and unaccountable. There is no logical reason why it should not have such a place; the laws of transactions have such a place in our account. But it is always intellectually satisfying when we can reduce the number of inexplicabilities. So far as I can discover, there is no good ground for supposing that motion is continuous; it is therefore worth while to develop a discontinuous hypothesis if we can thereby increase the unity and diminish the arbitrariness in our account of the physical world.

CHAPTER XXXVI

THE GENESIS OF SPACE-TIME

SPACE TIME, as it appears in mathematical physics, is obviously an artefact, *i.e.* a structure in which materials found in the world are compounded in such a manner as to be convenient for the mathematician. In the present chapter, I wish to collect what has already been said on this subject in various parts of the book, and to consider the resulting metaphysical status of space-time.

In the general theory of relativity, space-time appears in two ways: first, as providing a four-dimensional order; secondly, as giving rise to the metrical concept of "interval." Both are relations between "points," but both are treated mathematically as differential relations. This requires us to solve a purely mathematical problem: what is the function or process which tends towards these relations as a limit? This is on the assumption that space-time is continuous, which we do not know it to be. Let us begin with this hypothesis, and proceed afterwards to the hypothesis of discreteness. In the absence of evidence, it is necessary to develop both. For the present, therefore, I assume space-time to be continuous. This involves, or at least renders natural, the assumption that there is an infinite number of events compresent with any given event; I shall make this assumption also so long as I assume continuity.

"Compresence" is assumed to be a symmetrical relation, which every term in its field has to itself, and whose field is capable of being well ordered. A group of five events is capable of a relation called "co-punctuality," which means, in effect, that there is a region common to all five. A group of more than five events is called "co-punctual" when every quintet chosen out of it is co-punctual. A "point" is defined as a co-punctual group of events which cannot be added to without ceasing to be co-punctual. "Events" are defined as the field of the relation of compresence. Hence, by means of not unplausible axioms, we arrive at the space-time order presupposed in the assignment of co-ordinates. This part of the theory is straightforward.

When we come to "interval" there is more difficulty. In the discussion of measurement we decided, following Eddington, that equality of two intervals is what has to be defined, and that this has to be defined as a limit when both intervals tend towards zero. For this purpose, we supposed a relation of five points a, b, c, d, d', which we may express in the words: "*abcd'* is more nearly a parallelogram than *abcd*." From this, by means of a certain apparatus of axioms, we can arrive at what seems to be metrically necessary for mathematical physics. But this procedure is somewhat artificial. It seems natural to suppose that our relation of five points arises as follows: between any two points there is a relation, which for the moment we will call "separation," and the separation of a and b is more like that of c and d' than like that of c and d. Thus we shall have to do with degrees of resemblance between separations of point-pairs; these separations, however, cannot exist only for infinitesimal distances, but must exist for finite distances, at any rate if they are sufficiently small.

We have therefore to ask ourselves whether any physical meaning can be found for "separation," remembering that in the limit it is to have the properties of a small interval ds. This means to say that a separation may be of two sorts, space-like and time-like; also that the separation between two parts of a light-ray is zero. Now the separation will be time-like if there is any event at the one point which is a causal ancestor of an event at the other point; and the separation will be space-like if some event at the one point but not at the other and some event at the other but not at the one have a common ancestor or descendant, but no event at either is an ancestor or descendant of, or identical with, any event at the other. We shall assume that every pair of points has *some* causal relation, direct or indirect; that is to say, given any two events, e_1 and e_2, there will be somewhere in space-time two compresent events of which one is an ancestor or descendant of e_1 and the other of e_2. This is hardly more than a definition of the "world of physics"; for if an event had no causal relation, however indirect, to the part of the world which we know, it could never be inferred by us, and would in effect belong to a different universe. It follows that if two diverse points have neither a time-like nor a space-like separation, there is an event which is a member of both, but nothing at either is an effect of anything at the other. This happens with parts of a light-ray, if we suppose, as we have done, that it consists of steady events which persist until the light-ray is transformed into some other form of energy.

Thus we are led to the view that the relation of separation is somehow connected with the amount of causal action intervening between the two points concerned. It is easy to give a precise meaning to this idea when we assume a discrete space-time, but it is much more difficult in a continuous space-time. Nevertheless, it is perhaps not impossible.

Causality, for these purposes, may be confined to rhythms and transactions; mere relative motion, whether accelerated or uniform, will be regarded as not involving causality in the sense in which we mean it. Indirectly causality will be involved, since there will be a change of space-like separation; but the causality will be primarily concerned with other events, not with those constituting the biographies of the bodies in relative motion. In saying this, we are, I think, only interpreting the Einsteinian theory of gravitation.

In the preceding chapter, when we were considering a discrete space-time, we defined a time-like interval as the number of intervening points on the longest causal route connecting the two given points. The natural way to generalize this so as to become applicable to a continuous space-time would be to regard the number of points as the measure of geodesic distance; this would enable us to say that the geodesic distance traversed by a unit of matter measures the amount of causal action which it has undergone. If we further assume that, in comparing different units of matter, we must multiply by the mass to obtain a measure of the amount of causal action, then the amount in a finite motion is the integral of mds. But this is the amount of " action " in the technical sense.*

It seems therefore—though this is only a tentative suggestion—that we can regard a time-like separation as the measure of the maximum amount of causal action on the various causal routes which lead from one point to another. It is to be observed that, since points are classes of events, motion from one point to another consists in the cessation of certain events and the coming into existence of others; every such change is causal when it happens along the route of a piece of matter, since the unity of a piece of matter at different times is defined by means of the concept of a causal route. There is, therefore, so far as I can see, no fundamental objection to regarding time-like separations as measuring amounts of intervening causal action, and small time-like intervals as limits of separations. Space-like intervals, as we have seen, are derivative from time-like intervals; hence they, also, depend upon amount of causal action.

Passing now to the hypothesis of a discrete space-time, in which each point consists of a finite number of events, we find that a similar analysis to the above is still possible, and is in fact considerably easier than when we assume continuity.

* Eddington, *op. cit.*, p. 137.

In a discrete space-time, if P and P' are two points containing events which belong to the biography of one material unit, the number of points on the route of this unit between P and P' is always finite. If several geodesic routes lead from P to P', there will be a maximum to the number of points on such routes; this maximum will be the measure of the interval between P and P', which will therefore always be an integer. A longer route means a greater number of intermediate events, and therefore a greater amount of causal action. Thus again the interval measures the greatest amount of causal action on any causal route from P to P'. And causal routes consist of a succession of rhythms or steady events separated by transactions.

It will be observed that, in our theory, spatial distance does not directly represent any physical fact, but is a rather complicated way of speaking about the possibility of a common causal ancestry or posterity. For example, while a light-wave is supposed to be travelling away from an atom, it has no physical relation to anything in the atom subsequent to its emission. It may be reflected back to the atom after reaching some other atom, and then half the time of the double journey (as measured at the first atom) is called the spatial distance between the two atoms (taking the velocity of light as unity). But there is no adequate ground for asserting that at every moment of the intervening time the light-ray is at a certain spatial distance from the atom; indeed, the theory of relativity vetoes such a suggestion. There is therefore, so far as I can see, no reason in physics for believing in continuous motion, except as a convenient symbolic device for dealing with the time-relations of various discontinuous changes. And whether we regard space-time as continuous or discontinuous, motion loses its fundamental character, being replaced by successions of events belonging to the biographies of bits of matter. This is inevitable if we are to hold that motion is relative and action at a distance is a fiction.

There remains a question which is of some interest. Can time be derived from causality, or must we retain temporal order as fundamental, and distinguish cause and effect as the earlier and later terms in a causal relation ?* This question is bound up with that as to the reversibility of physical processes. If causal relations are symmetrical, so that whenever *A* and *B* are related as cause and effect it is physically possible that, on another occasion, *B* and *A* may be so related, then we must regard the time-order as something additional to the causal relation, not derivative from it. If, on the other hand, causal laws are irreversible, then we can define the time-order in terms of them, and need not introduce it as a logically separate factor. The question of reversibility is still *sub judice*, and I will not venture an opinion. The second law of thermodynamics asserts an irreversible process, but is purely statistical. All radiation of energy in spherical waves is *prima facie* irreversible, but we do not know that it really takes place. Dr Jeans suggests that there may also be converging spherical waves, and that these can be used to explain quantum phenomena.† For him, reversibility is a fundamental postulate.‡ I do not know whether he would maintain that the ejection of an electron or helium nucleus from a radio-active atom is a reversible process; but it must be confessed that, if it is not, the existence of radio-active elements becomes a mystery. Quantum theory has, on the whole, increased the arguments in favour of reversibility; but it cannot be said that there is as yet conclusive evidence on either side. We must, therefore, leave open the question whether the time-order of events in one causal route can be defined in terms of causal laws.

* This question (as well as various others) is ably discussed in a valuable article by Hans Reichenbach, *Kausalstruktur der Welt und der Unterschied von Vergangenheit und Zukunft*, Sitzungsberichte der Baierischen Akademie der Wissenschaften, mathematisch-naturwissenschaftliche Abteilung, 1925, pp. 133-175.

† *Op. cit.*, pp. 52-3.

‡ *Ib.*, p. 33.

CHAPTER XXXVII

PHYSICS AND NEUTRAL MONISM

In this chapter, I wish to define the outcome of our analysis in regard to the old controversy between materialism and idealism, and to make it clear wherein our theory differs from both. So long as the views set forth in previous chapters are supposed to be either materialistic or idealistic, they will seem to involve inconsistencies, since some seem to tend in the one direction, some in the other. For example, when I say that my percepts are in my head, I shall be thought materialistic; when I say that my head consists of my percepts and other similar events, I shall be thought idealistic. Yet the former statement is a logical consequence of the latter.

Both materialism and idealism have been guilty, unconsciously and in spite of explicit disavowals, of a confusion in their imaginative picture of matter. They have thought of the matter in the external world as being represented by their percepts when they see and touch, whereas these percepts are really part of the matter of the percipient's brain. By examining our percepts it is possible—so I have contended—to infer certain formal mathematical properties of external matter, though the inference is not demonstrative or certain. But by examining our percepts we obtain knowledge which is not purely formal as to the matter of our brains. This knowledge, it is true, is fragmentary, but so far as it goes it has merits surpassing those of the knowledge given by physics.

The usual view would be that by psychology we acquire knowledge of our "minds," but that the only way to acquire knowledge of our brains is to have them examined by a physiologist, usually after we are dead, which seems somewhat unsatisfactory. I should say that what the physiologist sees when he looks at a brain is part of his own brain, not part of the brain he is examining. The feeling of paradox about this view comes, I should say, from wrong views of space. It is true that what we see is not located where our percept of our own brain would be located if we could see our own brain; but this is a question of perceptual space, not of the space of physics. The space of physics is connected with causation in a manner which compels us to hold that our percepts are in our brains, if we accept the causal theory of perception, as I think we are bound to do. To say that two events have no spatio-temporal separation is to say that they are compresent; to say that they have a small separation is to say that they are connected by causal chains all of which are short. The percept must therefore be nearer to the sense-organ than to the physical object, nearer to the nerve than to the sense-organ, and nearer to the cerebral end of the nerve than to the other end. This is inevitable, unless we are going to say that the percept is not in space-time at all. It is usual to hold that "mental" events are in time but not in space; let us ask ourselves whether there is any ground for this view as regards percepts.

The question whether percepts are located in physical space is the same as the question of their causal connection with physical events. If they can be effects and causes of physical events, we are bound to give them a position in physical space-time in so far as interval is concerned, since interval was defined in causal terms. But the real question is as to "compresence" in the sense of Chapter XXVIII. Can a mental event be compresent with a physical event? If yes, then a mental event has a position in the space-time order; if no, then it has no such position. This, therefore, is the crucial question.

When I maintain that a percept and a physical event can

be compresent, I am not maintaining that a percept can have to a piece of matter the sort of relation which another piece of matter would have. The relation of compresence is between a percept and a physical *event*, and physical events are not to be confounded with pieces of matter. A piece of matter is a logical structure composed of events; the causal laws of the events concerned, and the abstract logical properties of their spatio-temporal relations, are more or less known, but their intrinsic character is not known. Percepts fit into the same causal scheme as physical events, and are not known to have any intrinsic character which physical events cannot have, since we do not know of any intrinsic character which could be incompatible with the logical properties that physics assigns to physical events. There is therefore no ground for the view that percepts cannot be physical events, or for supposing that they are never compresent with other physical events.

The fact that mental events admittedly have temporal relations has much force, now that time and space are so much less distinct than they were. It has become difficult to hold that mental events, though in time, are not in space. The fact that their relations to each other can be viewed as only temporal is a fact which they share with any set of events forming the biography of one piece of matter. Relatively to axes moving with the percipient's brain, the interval between two percepts of his which are not compresent should always be temporal, if his percepts are in his head. But the interval between simultaneous percepts of different percipients is of a different kind; and their whole causal environment is such as to make us call this interval space-like. I conclude, then, that there is no good ground for excluding percepts from the physical world, but several strong reasons for including them. The difficulties that have been supposed to stand in the way seem to me to be entirely due to wrong views as to the physical world, and more particularly as to physical space. The wrong views as to physical space have been encouraged by the notion that the primary qualities are objective, which has been held imaginatively by many men who would have emphatically repudiated it so far as their explicit thought was concerned.

I hold, therefore, that two simultaneous percepts of one percipient have the relation of compresence out of which spatio-temporal order arises. It is almost irresistible to go a step further, and say that any two simultaneous perceived contents of a mind are compresent, so that all our conscious mental states are in our heads. I see as little reason against this extension as against the view that percepts can be compresent. A percept differs from another mental state, I should say, only in the nature of its causal relation to an external stimulus. Some relation of this kind no doubt always exists, but with other mental states the relation may be more indirect, or may be only to some state of the body, more particularly the brain. "Unconscious" mental states will be events compresent with certain other mental states, but not having those effects which constitute what is called awareness of a mental state. However, I have no wish to go further into psychology than is necessary, and I will pursue this topic no longer, but return to matters of more concern to physics.

The point which concerns the philosophy of matter is that the events out of which we have been constructing the physical world are very different from matter as traditionally conceived. Matter was expected to be impenetrable and indestructible. The matter that we construct is impenetrable as a result of definition: the matter in a place is all the events that are there, and consequently no other event or piece of matter can be there. This is a tautology, not a physical fact; one might as well argue that London is impenetrable because nobody can live in it except one of its inhabitants. Indestructibility, on the other hand, is an empirical property, believed to be approximately but not exactly possessed by matter. I mean by indestructibility, not conservation of mass, which is known to be only approximate, but conservation of electrons and protons. At present it is not known whether an electron and a proton sometimes enter into a suicide pact or not,* but there is certainly no known reason why electrons and protons should be indestructible.

Electrons and protons, however, are not the stuff of the physical world: they are elaborate logical structures composed of events, and ultimately of particulars, in the sense of Chapter XXVII. As to what the events are that compose the physical world, they are, in the first place, percepts, and then whatever can be inferred from percepts by the methods considered in Part II. But on various inferential grounds we are led to the view that a percept in which we cannot perceive a structure nevertheless often has a structure, *i.e.* that the apparently simple is often complex. We cannot therefore treat the *minimum visible* as a particular, for both physical and psychological facts may lead us to attribute a structure to it—not merely a structure in general, but such and such a structure.

Events are neither impenetrable nor indestructible. Space-time is constructed by means of co-punctuality, which is the same thing as spatio-temporal interpenetration. Perhaps it is not unnecessary to explain that spatio-temporal interpenetration is quite a different thing from logical interpenetration, though it may be suspected that some philosophers have been led to favour the latter as a result of the arguments for the former. We are accustomed to imagining that numerical diversity involves spatio-temporal separation; hence we tend to think that, if two diverse entities are in one place, they

* It is thought highly probable that they do. See Dr Jeans, "Recent Developments of Cosmical Physics," *Nature*, December 4, 1926.

cannot be wholly diverse, but must be also in some sense one. It is this combination that is supposed to constitute logical interpenetration. For my part, I do not think that logical interpenetration can be defined without obvious self-contradiction; Bergson, who advocates it, does not define it. The only author I know of who has dealt seriously with its difficulties is Bradley, in whom, quite consistently, it led to a thorough-going monism, combined with the avowal that, in the end, all truth is self-contradictory. I should myself regard this latter result as a refutation of the logic from which it follows. Therefore, while I respect Bradley more than any other advocate of interpenetration, he seems to me, in virtue of his ability, to have done more than any other philosopher to disprove the kind of system which he advocated. However that may be, the spatio-temporal interpenetration which is used in constructing space-time order is quite different from logical interpenetration. Philosophers have been slaves of space and time in the imaginative application of their logic. This is partly due to Euler's diagrams and the notion that the traditional *A*, *E*, *I*, *O* were elementary forms of propositions and the confounding of "x is a β" with "all α's are β's." All this led to a confusion between classes and individuals, and to the inference that individuals can interpenetrate because classes can overlap. I do not suggest explicit confusions of this sort, but only that traditional elementary logic, taught in youth, is an almost fatal barrier to clear thinking in later years, unless much time is spent in acquiring a new technique.

On the question of the material out of which the physical world is constructed, the views advocated in this volume have, perhaps, more affinity with idealism than with materialism. What are called "mental" events, if we have been right, are part of the material of the physical world, and what is in our heads is the mind (with additions) rather than what the physiologist sees through his microscope. It is true that we have not suggested that all reality is mental. The positive arguments in favour of such a view, whether Berkeleyan or German, appear to me fallacious. The sceptical argument of the phenomenalists, that, whatever else there may be, we cannot know it, is much more worthy of respect. There are, in fact, if we have been right, three grades of certainty. The highest grade belongs to my own percepts; the second grade to the percepts of other people; the third to events which are not percepts of anybody. It is to be observed, however, that the second grade belongs only to the percepts of those who can communicate with me, directly or indirectly, and of those who are known to be closely analogous to people who can communicate with me. The percepts of minds, if such there be, which are not related to mine by communication—*e.g.* minds in other planets—can have, at best, only the third grade of certainty, that, namely, which belongs to the apparently lifeless physical world.

The events which are not perceived by any person who can communicate with me, supposing they have been rightly inferred, have a causal connection with percepts, and are inferred by means of this connection. Much is known about their structure, but nothing about their quality.

While, on the question of the stuff of the world, the theory of the foregoing pages has certain affinities with idealism—namely, that mental events are part of that stuff, and that the rest of the stuff resembles them more than it resembles traditional billiard-balls—the position advocated as regards scientific laws has more affinity with materialism than with idealism. Inference from one event to another, where possible, seems only to acquire exactness when it can be stated in terms of the laws of physics. There are psychological laws, physiological laws, and chemical laws, which cannot at present be reduced to physical laws. But none of them is exact and without exceptions; they state tendencies and averages rather than mathematical laws governing minimum events. Take, for example, the psychological laws of memory. We cannot say: At 12.55 G.M.T. on such and such a day, *A* will remember the event *e*—unless, indeed, we are in a position to remind him of it at that moment. The known laws of memory belong to an early stage of science—earlier than Kepler's laws or Boyle's law. We can say that, if *A* and *B* have been experienced together, the recurrence of *A* *tends* to cause a recollection of *B*, but we cannot say that it is sure to do so, or that it will do so in one assignable class of cases and not in another. One supposes that, to obtain an exact causal theory of memory, it would be necessary to know more about the structure of the brain. The ideal to be aimed at would be something like the physical explanation of fluorescence, which is a phenomenon in many ways analogous to memory. So far as causal laws go, therefore, physics seems to be supreme among the sciences, not only as against other sciences of matter, but also as against the sciences that deal with life and mind.

There is, however, one important limitation to this. We need to know in what physical circumstances such-and-such a percept will arise, and we must not neglect the more intimate qualitative knowledge which we possess concerning mental events. There will thus remain a certain sphere which will be outside physics. To take a simple instance: physics might, ideally, be able to predict that at such a time my eye would receive a stimulus of a certain sort; it might be able to trace the physical properties of the resulting events in the eye and the brain, one of which is, in fact, a visual percept; but it could not itself give us the knowledge that one of them is a visual percept. It is obvious that a man who can see knows things which a blind man cannot know; but a blind man can know the whole of physics. Thus the knowledge which other men have and he has not is not part of physics.

Although there is thus a sphere excluded from physics,

yet physics, together with a "dictionary," gives, apparently, all causal knowledge. One supposes that, given the physical characteristics of the events in my head, the "dictionary" gives the "mental" events in my head. This is by no means a matter of course. The whole of the foregoing theory of physics might be true without entailing this consequence. So far as physics can show, it might be possible for different groups of events having the same structure to have the same part in causal series. That is to say, given the physical causal laws, and given enough knowledge of an initial group of events to determine the purely physical properties of their effects, it might nevertheless be the case that these effects could be qualitatively of different sorts. If that were so, physical determinism would not entail psychological determinism, since, given two percepts of identical structure but diverse quality, we could not tell which would result from a stimulus known only as to its physical, *i.e.* structural, properties. This is an unavoidable consequence of the abstractness of physics. If physics is concerned only with structure, it cannot, *per se*, warrant inferences to any but the structural properties of events. Now it may be a fact that (*e.g.*) the structure of visual percepts is very different from that of tactual percepts; but I do not think such differences could be established with sufficient strictness and generality to enable us to say that such-and-such a stimulus must produce a visual percept, while such another must produce a tactual percept.

On this matter, we must, I think, appeal to evidence which is partly psychological. We do know, as a matter of fact, that we can, in normal circumstances, more or less infer the percept from the stimulus. If this were not the case, speaking and writing would be useless. When the lessons are read, the congregation can follow the words in their own Bibles. The differences in their "thoughts" meanwhile can be connected causally, at least in part, with differences in their past experience, and these are supposed to make themselves effective by causing differences in the structure of brains. All this seems sufficiently probable to be worth taking seriously; but it lies outside physics, and does not follow from the causal autonomy of physics, supposing this to be established even for human bodies. It will be observed that what we are now considering is the converse of what is required for the inference from perception to physics. What is wanted there is that, given the percept, we should be able to infer, at least partially, the structure of the stimulus—or at any rate that this should be possible when a sufficient number of percepts are given. What we want now is that, given the *structure* of the stimulus (which is all that physics can give), we should be able to infer the *quality* of the percept—with the same limitations as before. Whether this is the case or not, is a question lying outside physics; but there is reason to think that it is the case.

The aim of physics, consciously or unconsciously, has always been to discover what we may call the causal skeleton of the world. It is perhaps surprising that there should be such a skeleton, but physics seems to prove that there is, particularly when taken in conjunction with the evidence that percepts are determined by the physical character of their stimuli. There is reason—though not quite conclusive reason—for regarding physics as causally dominant, in the sense that, given the physical structure of the world, the qualities of its events, in so far as we are acquainted with them, can be inferred by means of correlations. We have thus in effect a psycho-cerebral parallelism, although the interpretation to be put upon it is not the usual one. We suppose that, given sufficient knowledge, we could infer the qualities of the events in our heads from their physical properties. This is what is really meant when it is said, loosely, that the state of the mind can be inferred from the state of the brain. Although

I think that this is probably true, I am less anxious to assert it than to assert, what seems to me much more certain, that its truth does not follow from the causal autonomy of physics or from physical determinism as applied to all matter, including that of living bodies. This latter result flows from the abstractness of physics, and belongs to the philosophy of physics. The other proposition, if true, cannot be established by considering physics alone, but only by a study of percepts for their own sakes, which belongs to psychology. Physics studies percepts only in their cognitive aspect; their other aspects lie outside its purview.

Even if we reject the view that the quality of events in our heads can be inferred from their structure, the view that physical determinism applies to human bodies brings us very near to what is most disliked in materialism. Physics may be unable to tell us what we shall hear or see or " think," but it can, on the view advocated in these pages, tell us what we shall say or write, where we shall go, whether we shall commit murder or theft, and so on. For all these are bodily movements, and thus come within the scope of physical laws. We are often asked to concede that the beauties of poetry or music cannot result from physical laws. I should concede that the beauty does not result from physics, since beauty depends in part upon intrinsic quality; if it were, as some writers on æsthetics contend, solely a matter of form, it would come within the scope of physics, but I think these writers do not realize what an abstract affair form really is. I should concede also that the *thoughts* of Shakespeare or Bach do not come within the scope of physics. But their thoughts are of no importance to us: their whole social efficacy depended upon certain black marks which they made on white paper. Now there seems no reason to suppose that physics does not apply to the making of these marks, which was a movement of matter just as truly as the revolution of the earth in its orbit. In any case, it is undeniable that the socially important part of their thought had a one-one relation to certain purely physical events, namely the occurrence of the black marks on the white paper. And no one can doubt that the causes of our emotions when we read Shakespeare or hear Bach are purely physical. Thus we cannot escape from the universality of physical causation.

This, however, is perhaps not quite the last word on the subject. We have seen that, on the basis of physics itself, there may be limits to physical determinism. We know of no laws as to when a quantum transaction will take place or a radio-active atom will break down. We know fairly well what will happen *if* anything happens, and we know statistical averages, which suffice to determine macroscopic phenomena. But if mind and brain are causally interconnected, very small cerebral differences must be correlated with noticeable mental differences. Thus we are perhaps forced to descend into the region of quantum transactions, and to desert the macroscopic level where statistical averages obtain. Perhaps the electron jumps when it likes; perhaps the minute phenomena in the brain which make all the difference to mental phenomena belong to the region where physical laws no longer determine definitely what must happen. This, of course, is merely a speculative possibility; but it interposes a veto upon materialistic dogmatism. It may be that the progress of physics will decide the matter one way or other; for the present, as in so many other matters, the philosopher must be content to await the progress of science.

CHAPTER XXXVIII

SUMMARY AND CONCLUSION

In the present state of physics, many questions of considerable philosophical importance cannot be answered, although they are such as science may hope to answer, and largely such as were formerly supposed to have been already answered. This makes the task of the philosopher more difficult; it is necessary to develop various hypotheses, so as to be prepared for whatever decision science may arrive at. Certain things, it is true, may be taken as definitely ascertained; these things, so far as they are relevant to philosophy, were considered in Part I. It is clear that, in some sense, there are electrons and protons, and we cannot well doubt the substantial accuracy of their estimated masses and electric charge. That is to say, these constants evidently represent something of importance in the physical world, though it would be rash to say that they represent exactly what is at present supposed. In like manner there seems to be no reasonable doubt that there is a constant h, whose dimensions are those of action or angular momentum, and whose magnitude is substantially what it has been estimated to be. It would seem clear also that h is a constant which is characteristic of periodic processes. Moreover, the change from one such process to another, which is what we have called a transaction, is governed by principles connected with h in addition to the conservation of energy.

But it would be very rash to maintain that the current mathematical formulation of the quantum principle is the best possible; indeed, there are reasons for dissatisfaction with it. Perhaps the most important of these is that in expressing the kinetic energy we have to employ the method of separation of variables, and that we do not know whether separation of variables is always possible, or whether all ways of separating the variables give equivalent results. Apart from these rather technical difficulties, there are others that are less definite but perhaps not less important. No one has succeeded in making the existence of quanta seem at all "reasonable"; that is to say, it remains isolated and separate from other physical ideas. And whereas it involves discontinuity, the whole effect of relativity has been to emphasize continuity. Moreover, no one has yet succeeded in explaining interference and diffraction by means of light-quanta, or in explaining the photo-electric effect without them. For these reasons, the time has not yet come when the philosopher can deal confidently with quantum theory; he can only suggest what would be his philosophy if this or that view had prevailed in physics.

In relativity, we are on surer ground. The advance on the physics of the past, where relativity is concerned, is mainly logical and philosophical. It is true that facts led to the theory, and that the theory in turn led to the discovery of new facts. But the facts were small and only just within the limits of observation; and they had not, as facts, the revolutionary importance of the facts about quanta. And now that the theory is fairly complete, one can see that, theoretically, it ought to have been discovered by Galileo, or at any rate as soon as the velocity of light became known. It represents in its technique a better philosophy than that of Newton; indeed, one of its most remarkable features is the adaptation of the technique to the philosophy.

The theory of relativity, to my mind, is most remarkable when considered as a logical deductive system. That is the reason, or one of the reasons, why I have found occasion to allude so constantly to Eddington. He, more than Einstein or Weyl, has expounded the theory in the form most apt for the purposes of the philosopher. Minkowski had the same quality, but he did not live to see the general theory. For philosophical purposes, therefore, I have allowed myself to be guided almost entirely by Eddington.

In the general theory of relativity, we start with a four-dimensional continuum of points, whose properties, to begin with, are purely ordinal. We then assign four co-ordinates to each point on any principle such that the ordinal properties of the co-ordinates are the same as those of the points. We then assume that, if two points are very close together, there is a quadratic function of the co-ordinates which has the same value however the co-ordinates may be assigned, subject to the above ordinal condition. If this function is positive, its square root is called the (time-like) interval; if negative, the square root of the function with its sign changed is called the (space-like) interval. Omitting niceties, we may say that the remainder of the theory turns mainly on geodesics. A geodesic is a route between two space-time points such that the integral of the interval along this route is stationary. In the important routes, it is a maximum. It appears that energy can be divided into parcels which move in geodesics; when these parcels move with a velocity less than that of light, they are regarded as pieces of matter. Weyl, by imposing certain limitations on measurement, succeeds in including electro-magnetic phenomena in this scheme. Thus we have a comprehensive theory which may be taken to embrace everything except quantum phenomena.

But although there is so much to give pleasure to the logician in this scheme—more especially the method of tensors and Hamiltonian derivatives—yet the philosopher cannot but feel dissatisfaction with the apparently arbitrary assumption about intervals. This assumption seemed less arbitrary than it is, because of its connection, historically, with the theorem of Pythagoras and its modifications in non-Euclidean geometry. But the theorem was believed formerly because it had been proved; when the proof was found to have no value, it was believed because empirical evidence was thought to show its approximate truth. This empirical evidence, of course, remains, but the theory of relativity has made its value much more problematical than it formerly seemed. And it is customary to carry out measurements carefully, taking trouble to secure bodies that are as nearly rigid as possible, and optical instruments that are accurate. If our co-ordinates are to be arbitrary, as they are in the general theory of relativity, it is doubtful whether we still have a right to expect that they will verify anything analogous to the theorem of Pythagoras.

As against these doubts, it may be said that the general theory has justified itself by the correctness of all its verifiable consequences. This is true, and I do not wish to minimize the force of the argument. But I seem to observe that, in obtaining these results, the theory does not make use of the full liberty in assignment of co-ordinates which it claims at the start. In astronomy, its co-ordinates are still assigned by the usual careful methods, and it is not clear that this care is useless. From the method of tensors, it *seems* to follow that we can employ any co-ordinates subject to the ordinal condition. But the method of tensors, as used, assumes the formula for interval; for this reason, Dr Whitehead found it necessary, in his *Principle of Relativity*, to give a theory of tensors independent of the formula for interval. There is thus still legitimate room for doubt as to whether the formula for interval is really quite independent of the choice of co-ordinates.

And, apart from this question, there is great difficulty in suggesting any non-technical meaning for interval; yet such a meaning ought to exist, if interval is as fundamental as it appears to be in the theory of relativity. There is difficulty also as to what is meant by measurement. And there is the feeling that, perhaps, tensor equations represent purely ordinal properties of the space-time continuum, and could, by a better technique, be set forth without the use of any co-ordinates at all. The necessary technique does not exist at present, but it is not impossible that it may be created before long.

In Part II., we approached a different type of question: the question of the evidence for the truth of physics, *i.e.* of the relation of physics to perception. For the purposes of this inquiry, it is convenient to use "perception" somewhat more narrowly than it would be used in psychology. Our purpose is epistemological, and therefore perception is only relevant in so far as it is explicit and the percept is observed: percepts which pass unnoticed cannot be made into premisses for physics. The use of percepts for inference as to the physical world rests upon the causal theory of perception, since the naive realism of common sense turns out to be self-contradictory. The serious alternatives to the causal theory of perception are not common sense, but solipsism and phenomenalism. Solipsism, as an epistemologically serious theory, must mean the view that from the events which I experience there is no valid method of inferring the character, or even the existence, of events which I do not experience. If inference is taken in the sense of strict deductive logic, there is, so far as I can see, no escape from the solipsist position. And it should be observed that this position cannot admit unconscious events in me, any more than events outside me: its basis is epistemological, and therefore, for it, the important distinction is between what I experience and what I do not experience, not between what is mine and what is not mine in some metaphysical or physical sense. We cannot escape from the solipsist position without bringing in induction and causality, which are still subject to the doubts resulting from Hume's sceptical criticism.

Since, however, all science rests upon induction and causality, it seems justifiable, at least pragmatically, to assume that, when properly employed, they can give at least a probability. In the present work, I have made this assumption baldly, without attempting to justify it; I have done this because I do not believe that a justification could be much briefer than Mr Keynes's *Treatise on Probability*, and also because, while I am convinced that a justification is possible, I am not satisfied with those put forward by others or with any that I have been able to invent myself. It seemed best, therefore, to make the assumption as stark as possible, without any attempt at artificial plausibility.

Intermediate between solipsism and the ordinary scientific view, there is a half-way house called "phenomenalism." This admits events other than those which I experience, but holds that all of them are percepts or other mental events. Practically, it means, when advocated by scientific men, that they will accept the testimony of other observers as to what they have actually experienced, but that they will not infer thence anything which no observer has experienced. It may be said, in justification of this position, that, while it employs analogy and induction, it refrains from assuming causality. But it may be doubted whether it can really abstain from causality. Phenomenalists appear to take testimony for granted, *i.e.* to assume that the words which they see and hear express what they themselves would express if they used them. But this involves causality, and involves it in the form in which the cause is in one person and the effect in another. There does not seem, therefore, to be any substantial justification for this half-way house.

We therefore assume, though with less than demonstrative certainty, that percepts have causes which may be not percepts, and, in particular, that when a number of people have similar percepts simultaneously, there is what may be called a "field" of causally connected events, which, it is found, have relations that often enable us to arrange them in a spherical order about a centre. We thus arrive at a space-time order of events, which is found to be the same whichever of many possible methods of arriving at it we adopt; in this order, a percept is located in the head of the percipient. In drawing inferences from percepts to their causes, we assume that the stimulus must possess whatever structure is possessed by the percept, though it may also have structural properties not possessed by the percept. The assumption that the structural properties of the percept must exist in the stimulus follows from the maxim "same cause, same effect" in the inverted form "different effects, different causes," from which it follows that if, *e.g.*, we see red and green side by side, there is some difference between the stimulus to the red percept and the stimulus to the green percept. The structural features possessed by the stimulus but not by the percept, when they can be inferred, are inferred by means of general laws—*e.g.* when two objects look similar to the naked eye but dissimilar under the microscope, we assume that there are differences in the stimuli to the naked-eye percepts which produce either no differences, or no perceptible differences, in the corresponding percepts.

These principles enable us to infer a great deal as to the structure of the physical world, but not as to its intrinsic character. They put percepts in their place as occurrences analogous to and connected with other events in the physical world, and they enable us to regard a dictaphone or a photographic plate as having something which, from the standpoint of physics, is not very dissimilar from perception. We no longer have to contend with what used to seem mysterious in the causal theory of perception: a series of light-waves or sound-waves or what not suddenly producing a mental event apparently totally different from themselves in character. As to intrinsic character, we do not know enough about it in the physical world to have a right to say that it is very different from that of percepts; while as to structure we have reason to hold that it is similar in the stimulus and the percept. This has become possible owing to the facts that "matter" can be regarded as a system of events, not as part of the stuff of the world, and that space-time, as it occurs in physics, has been found to be much more different from perceptual space than was formerly imagined.

This brings us to Part III., in which we endeavour to discover a possible structure of the physical world which shall at once justify physics and take account of the connection with perception demanded by the necessity for an empirical basis for physics. Here we are concerned first with the construction of points as systems of events which overlap, or are "co-punctual," in space-time, and then with the purely ordinal properties of space-time. The method employed is very general, and can be adapted to a discrete or to a continuous order; it is proved that $\aleph_0$ events are sufficient to generate a continuum of points, given certain laws as to the manner of their overlapping. The whole of this theory, however, aims only at constructing such properties of space-time as belong to *analysis situs;* everything appertaining to intervals and metrics is omitted at this stage, since causal considerations are required for the theory of intervals.

The conception of one unit of matter—say one electron—as a "substance," *i.e.* a single simple entity persisting through time, is not one which we are justified in adopting, since we have no evidence whatever as to whether it is false or true. We define a single material unit as a "causal line," *i.e.* as a series of events connected with each other by an intrinsic differential causal law which determines first-order changes, leaving second-order changes to be determined by extrinsic causal laws. (In this we are for the moment ignoring quantum phenomena.) If there are light-quanta, these will more or less fulfil this definition of matter, and we shall have returned to a corpuscular theory of light; but this is at present an open question. The whole conception of matter is less fundamental to physics than it used to be, since energy has more and more taken its place. We find that under terrestrial conditions electrons and protons persist, but there is nothing in theoretical physics to lead us to expect this, and physicists are quite prepared to find that matter can be annihilated. This view is even put forward to account for the energy of the stars.

The question of interval presents great difficulties, when we attempt to construct a picture of the world which shall make its importance seem not surprising. The same may be said of the quantum. I have endeavoured, not, I fear, with much success, to suggest hypotheses which would link these two curious facts into one whole. I suggest that the world consists of steady events accompanied by rhythms, like a long note on the violin while arpeggios are played on the piano, or of rhythms alone. Steady events are of various sorts, and many sorts have their appropriate rhythmic accompaniments. Quantum changes consist of "transactions," *i.e.* of the substitution, suddenly, of one rhythm for another. When two events have a time-like interval, if space-time is discrete, this interval is the greatest number of transitions on any causal route leading from the one event to the other. The definition of space-like intervals is derived from that of time-like intervals. The whole process of nature may, so far as present evidence goes, be conceived as discontinuous; even the periodic rhythms may consist of a finite number of events per period. The periodic rhythms are required in order to give an account of the uses of the quantum principle. A percept, at any rate when it is visual, will be a steady event, or system of steady events, following upon a transaction. Percepts are the only part of the physical world that we know otherwise than abstractly. As regards the world in general, both physical and mental, everything that we know of its intrinsic character is derived from the mental side, and almost everything that we know of its causal laws is derived from the physical side. But from the standpoint of philosophy the distinction between physical and mental is superficial and unreal.

INDEX